Understanding User–Web Interactions via Web Analytics

Synthesis Lectures on Information Concepts, Retrieval, and Services

Editor
Gary Marchionini, University of North Carolina, Chapel Hill

Understanding User – Web Interactions via Web Analytics
Bernard J. (Jim) Jansen
2009

XML Retrieval
Mounia Lalmas
2009

Faceted Search
Daniel Tunkelang
2009

Introduction to Webometrics: Quantitative web research for the social sciences
Michael Thelwall
2009

Automated Metadata in Multimedia Information Systems: Creation, Refinement, Use in Surrogates, and Evaluation
Michael G. Christel
2009

Exploratory Search: Beyond the Query-Response Paradigm
Ryen W. White and Resa A. Roth
2009

New Concepts in Digital Reference
R. David Lankes
2009

Understanding User–Web Interactions via Web Analytics
Bernard J. (Jim) Jansen
www.morganclaypool.com

ISBN: 9781598298512 paperback

ISBN: 9781598298529 ebook

DOI: 10.2200/S00191ED1V01Y200904ICR006

A Publication in the Morgan & Claypool Publishers series

SYNTHESIS LECTURES ON INFORMATION CONCEPTS, RETRIEVAL, AND SERVICES #6

Series Editor: Gary Marchionini, University of North Carolina, Chapel Hill

Series ISSN

ISSN 1947-945X   print

ISSN 1947-9468   electronic

Understanding User–Web Interactions via Web Analytics

Bernard J. (Jim) Jansen
Pennsylvania State University

SYNTHESIS LECTURES ON INFORMATION CONCEPTS, RETRIEVAL, AND SERVICES #6

MORGAN & CLAYPOOL PUBLISHERS

ABSTRACT

This lecture presents an overview of the Web analytics process, with a focus on providing insight and actionable outcomes from collecting and analyzing Internet data. The lecture first provides an overview of Web analytics, providing in essence, a condensed version of the entire lecture. The lecture then outlines the theoretical and methodological foundations of Web analytics in order to make obvious the strengths and shortcomings of Web analytics as an approach. These foundational elements include the psychological basis in behaviorism and methodological underpinning of trace data as an empirical method. These foundational elements are illuminated further through a brief history of Web analytics from the original transaction log studies in the 1960s through the information science investigations of library systems to the focus on Websites, systems, and applications. Following a discussion of on-going interaction data within the clickstream created using log files and page tagging for analytics of Website and search logs, the lecture then presents a Web analytic process to convert these basic data to meaningful key performance indicators in order to measure likely converts that are tailored to the organizational goals or potential opportunities. Supplementary data collection techniques are addressed, including surveys and laboratory studies. The overall goal of this lecture is to provide implementable information and a methodology for understanding Web analytics in order to improve Web systems, increase customer satisfaction, and target revenue through effective analysis of user–Website interactions.

KEYWORDS

Web analytics, search log analysis, transaction log analysis, transaction logs, log file, query logs, key performance indicators, query log analysis, Web search research, Webometrics

Preface

I have based this lecture on my research and practical work in the Web analytics area, along with the work of many others. One advantage of sustained work over time in a given field is the ability to go back and correct, modify, expand, and improve, with any luck, previous efforts, documentations, and writings. Additionally, sustained contribution to a body of knowledge in an area often leads one to works by and interchanges with other researchers, practitioners, and scholars who enhance and add to the field. This lecture is the outcome of this continual and reiterative learning process. I hope the content within jumpstarts the learning process of others in the field of Web analytics.

My goal with this lecture is to present the conceptual aspects of Web analytics, relating these facets to processes and concepts that address the pertinence of Web analytics, and provide meaning to the techniques used in Web analytics. To that end, each section of the lecture represents a major component of the Web analytics field. While many sections are built on previous publications, I have enhanced each with updated thoughts and directions in the field, drawing from my own insights as well as those of others who are pushing and defining the field of Web analytics, including both academics and practitioners. Collectively, the sections offer an integrated and coherent primer of the exciting field of Web analytics.

This is not necessarily a "how-to" book. The field of Web analytics is dynamic, and any "how-to" book would likely be out of date before it could be published. Instead, in this lecture, I offer foundational elements that are more enduring than implementation techniques could be. As such, the lecture may have some enduring value.

Acknowledgments

I sincerely acknowledge and thank the many collaborators with whom I have worked with in the Web analytics area. I am appreciative of suggestions of material from Eric Peterson and Mark Ruzomberka, both excellent pracitioners in the art and science of Web analytics. I am also thankful to the reviewers of this manuscript for their valuable input, specifically Dietmar Wolfram and Fabrizio Silvestri, both excellent researchers in the field of Web log analysis. Finally, I am indebted to Diane Cerra and Gary Marchionini for their patience and assistance.

Contents

CHAPTER 1

Understanding Web Analytics

Let us pretend for the moment that we run an online retail store that sells a physical product, perhaps the latest athletic shoe, as just an example. How do potential customers find our online store? Do they find us via major search engines or from other sites? How will we know, and why should we care? What might it mean if they come to our Website and then immediately leave? What if the potential customer explores several pages and then leaves? Do these customers' actions tell us anything valuable about our Website or call for actions on our part? If a customer starts to make a purchase but then leaves before completing the order, should we look at a site redesign? To make our hypothetical online store successful, we need to understand why potential customers behave as they do, and the possible answers to our questions lie within the field of Web analytics.

The Web Analytics Association (WAA) defines Web analytics as "the measurement, collection, analysis, and reporting of Internet data for the purposes of understanding and optimizing Web usage" (http://www.webanalyticsassociation.org/).

This seemingly clear-cut definition is not so clear-cut when we consider the numerous unstated assumptions, methods, and tools needed for its implementation. In fact, the definition raises several critically unanswered questions. For example, the definition leaves Internet data undefined. What is this Internet data? What are its strengths and shortcomings? Where does one get it? Once defined, collection implies some application that can do the collecting. What application is doing the collecting? Measurement implies processes and benchmarks. Where are these processes and benchmarks? Analysis implies both a methodology and strategy for its conduct, which leads one from the data to understanding and insight. What is this methodology, and how does one define the strategy? What constructs is this strategy based upon? Reporting implies an organizational unit in which to report for some external purpose. Optimizing implies a focus on technology or processes. Understanding implies a focus on people or contexts.

Answering some of these questions is the goal of this lecture, and the questions and assumptions of our definition provide the structure for our discussion of the increasingly important field of Web analytics.

Given the commercial structure of the Web, Web analytics has typically taken a business perspective in the practitioner arena. Within academia, a near parallel movement is focusing on transaction log analysis (TLA) and Webometrics areas [147], where interesting academic research is occurring. In this lecture, the perspective will shift from one paradigm to the other. Because the jargon from the practitioner side is rapidly gaining wider acceptance, this lecture leverages that terminology, in the main. As such, much of the discussion will take a "business" and "customer" perspective that could be somewhat alien to some academic readers. I would recommend these academic readers to adapt. This is the direction that the field is heading, and in the main, it is for the better. It is where the action is. Practice is informing research.

The commercial force on the Web is pushing Web analytics research outside of academia at a near unbelievable pace. The drivers for these movements are clear. The Web has significantly shortened the distance between a business and its customers, both physically and emotionally. The distance between business and customer is now the duration of a single click. These clicks drive the economic models that support our Web search engines and provide the economic fuel for an increasing number of businesses. The click (with the associated customer behavior that accompanies it) is at the heart of an economic engine that is changing the nature of commerce with the near instantaneous, real-time recording of customer decisions to buy or not to buy (or some other analyst defined conversion other than purchase).

As such, Web analytics deals with Internet customer interaction data from Web systems. From an academic research perspective, this data is known as trace data (i.e., traces left behind that indicate human behaviors). A basic premise of Web analytics is that this user-generated Internet data can provide insight to understanding these users better or point to needed changes or improvements to existing Web systems. This can be data collected directly on a given Website or gathered indirectly from other applications. Almost all direct data that we can collect is behavioral data, which is data that relates to the behavior of a user on a Web system. As such, this data provides wonderful insights into what a user is doing. It tells us the "what." However, its shortcoming is that it offers little insight into the motivations or decision processes of that user. These are what academics call the contextual, situational, cognitive, and affective aspects of the user. For example, a click online could indicate extreme interest, slight consideration, or perhaps a serendipitous experience. To explore the "why," we need attitudinal data (i.e., the contextual, situational, cognitive, and affective stuff). For these insights, one must typically use other forms of data collection methods to supplement behavioral data, such as surveys, interviews, or laboratory studies. However, behavioral data is a great starting point to isolate the most promising possibilities (based on some external goal) and then move to attitudinal data collection methods in order to investigate possible meanings or solutions.

As researchers, we collect this behavioral data using an application that logs user behavior on the Website, along with other associated measures. These logging applications come in a variety of flavors, with continually changing structure, coding, and features. However, they all perform the same core activities—collect and archive data in some type of storage location. This storage location is typically a log file, characteristically known as a transaction log, hence the name *transaction log analysis* (i.e., Web analytics in academic circles). While transaction log formats vary, they all generally report similar behavioral data, along with associated contextual data concerning the computer and related software (i.e., operating system, browser type, etc.).

The issue of the computer is extremely important as it serves to highlight one of the key shortcomings of Internet data, namely, that the data can sometimes be inaccurate. There are several sources of data error. Primarily, Internet behavioral data are traceable back to a computer or computer browser and not necessarily to an exact person, assuming the person does not log into an account. In many cases this issue may make little difference. If a product is sold, a product is sold. However, in other situations such as search and visitor counts, this issue can cause numerous problems. This is especially so with Web search engines where one can log on anonymously. In addition, common use computers can skew the data. Additionally, with the proliferation of scrapping software one cannot always tell whether the visitor was even a human. In the case of popular Websites, many times most of the server load is software generated. The behavior of these bots can significantly skew the data. Other sources of data inaccuracies include the use of cookies, internal visitors, caching servers, and incorrect page tagging. Finally (and we computer scientists rarely discuss this), data catch applications are not perfect, based on personal experience error rates are often in the 5% to 10% range.

Once we have collected the data, as accurately as possible, we begin the process of getting value by reporting and analyzing it. Reporting is somewhat straightforward and generally involves compiling data in some aggregate way for clarity and simplification. In analysis, we attempt to leverage the data to understand some set goal and, perhaps, to make recommendations for improvement, identify opportunities, or highlight specific findings. In order to enable practitioners to get value from the analysis, researchers must establish proven processes and methodologies. Methodologies must be identified and used to correct for data inaccuracy and typically must be scalable (i.e., able to handle large volumes of data). The tactical aspects of analysis (and the related issues of data cleaning before analysis) can be extremely time-consuming until one establishes an efficient procedure. However, several commercial tools can aid in the process. The effective analysis must generate the proper metrics and key performance indicators (KPIs).

From a formal perspective KPIs measure performance based on articulated goals for the business, user understanding, or Web system. Each KPI, then, should link directly to goals; therefore, KPIs enable goal achievement by defining and measuring progress. The setting of these KPIs is of paramount importance in achieving a related technology, user, or organizational goal.

In defining KPIs, we identify the actions that are desired behaviors and then relate these desired behaviors toward measurable goals. KPIs will vary based on the organization and Web system. Typically, KPIs for commercial sites are overall purchase conversions, average order size, and items per order. For lead generation sites, KPIs might be overall conversions, conversion by campaigns, dropouts, and conversions of leads to actual customers. Customer service sites might focus on reducing expenses and improving customer experiences. For advertising on content sites, KPIs could be visits per week, page viewed per visit, visit length, advertising click ratio, and ratio of new to returning visitors.

Analysis results are of little value until one takes action driven by the data that is in line with the established KPI. One generally refers to this as actionable outcomes. In academic circles, this may mean generating publications that shed insight on user behavior, or changes to some methods or system. In a business, this means calculated change to improve the Website or business process that is directly dependent on the KPI selected.

We directly link KPIs to goals by monetizing (i.e., assigning value to) the desired behaviors that these indicators reflect. Generally, these goals relate to generating additional revenue, reducing costs, or improving the user experience. If we want more visitors, we must determine how much each visitor is worth to us. If we are interested in items ordered, we identify the value of each additional item ordered to the organization. By clearly articulating this linkage between KPIs and goals, we can then see the impact of these indicators and make choices about prioritizing opportunities and problems. This type of analysis can also aid in eliminating unsuccessful projects and determining the impact of system changes. Such a linkage process can aid in determining the value of Web campaigns and recognizing the investment return on Web system use.

In a nutshell, this is the field of Web analytics. In the following sections of this lecture, we investigate each of the concepts and areas in more detail, beginning with an examination of the theoretical foundations of Web analytics.

*   *   *   *

CHAPTER 2

The Foundations of Web Analytics: Theory and Methods

What are the foundational elements that provide confidence that Web analytics is providing useful insights? To address such a question, we must investigate the underlying constructs of Web analytics. This section explains the theoretical and methodological foundations for Web analytics, addressing the fundamentals of the field from a research viewpoint and the concept of Web logs as a data collection technique from the perspective of behaviorism. By behaviorism, we take a more liberal view than is traditional, as will be explained.

From this research foundation, we then move to the methodological aspects of Web analytics and examine the strengths and limitations of Web logs as trace data. We then review the conceptualization of Web analytics as an unobtrusive approach to research and present the power and deficiency of the unobtrusive methodological concept, including benefits and risks of Web analytics specifically from the perspective of an unobtrusive method. The section also highlights some of the ethical questions concerning the collection of data via Web log applications.

Conducting research involves the use of both a set of theoretical constructs and methods for investigation [74]. For empirical research, the results are linked conceptually to the data collection process. High-quality research requires a thorough methodological frame. In order to understand empirical research and the implications of the results, we must thoroughly understand the techniques by which the researcher collected and analyzed data. A variety of methods is available for research concerning users and information systems on the Web, including qualitative, quantitative, and mixed methods. The selection of an appropriate method is critical if the research is to have efficient execution and effective outcomes. The method of data collection also involves a choice of methods. Web logs (including both transaction logs and search logs) and Web analytics (including TLA and search log analysis [SLA]) are approaches to data collection and research methodology, respectively, for both system performance and user behavior analysis that has been used since 1967 [105], in peer-reviewed research since 1975 [116], and in numerous practitioner outlets since the 1990s [118].

A Web log is an electronic record of interactions that have occurred between a system and users of that system. These log files can come from a variety of computers and systems (Websites, to online public access catalogs or OPACs, user computers, blogs, listserv, online newspapers, etc.), basically any application that can record the user–system–information interactions. Web analytics also takes various forms but commonly involves TLA, which was preceded by log analysis in the academic fields of library, information, and computer science. TLA is the methodological approach to studying online systems and users of these systems. Peters [117] defines TLA as the study of electronically recorded interactions between online information retrieval systems and the persons who search for information found in those systems. Since the advent of the Internet, we have had to modify Peters' (1993) definition, expanding it to include systems other than information retrieval systems. In general, the practitioner side of Web analytics seems to have developed relatively independently, with few people venturing out and sharing learning between the practitioner and academic camps.

Partly as a result of this separate development, Web analytics is a broad categorization of methods that covers several sub-categorizations, including TLA (i.e., analysis of any log from a system), Web log analysis (i.e., analysis of Web system logs), blog analysis (i.e., analysis of Web logs), and SLA (analysis of search engine logs), among others. The study of digital libraries is also an interesting domain that involves both searching and browsing. Web analytics enables macro-analysis of aggregate user data and patterns and microanalysis of individual search patterns. The results from the analyzed data help to develop systems and services based on user behavior or system performance, and these services and performance enhancements are usually leveraged to achieve other goals.

From a user behavior perspective, Web analytics is one of a class of unobtrusive methods (a.k.a., non-reactive or low-constraint). Unobtrusive methods are those that allow data collection without directly contacting participants. The research literature specifically describes unobtrusive approaches as those that do not require a direct response from participants [102, 112, 154]. This data can be gathered through observation or from existing records. In contrast to unobtrusive methods, obtrusive or reactive approaches, such as questionnaires, tests, laboratory studies, and surveys, require a direct response from participants [153]. A laboratory experiment is an example of an extremely obtrusive method. The metaphorical line between unobtrusive and obtrusive methods is unquestionably blurred, and instead of one thin line there is a rather large gray area. For example, conducting a survey to gauge the reaction of users to information systems is an obtrusive method. However, using the posted results from the survey is an unobtrusive method. Granted, this may be making a strictly intellectual distinction, but the point is that log data falls in the gray area. In some respects, users know that their actions are being logged or recorded on Websites. However, logging applications are generally so unobtrusive that they fade into the background [4].

With this introduction, we now address the specific research and methodological foundations of Web analytics. We first address the concept of transaction logs as a data collection technique from the perspective of behaviorism, and then review the conceptualization of Web analytics as trace data and an unobtrusive method. We present the strengths and shortcomings of the unobtrusive methodology approach, including benefits and shortcomings of Web analytics specifically from the perspective of an unobtrusive method. We end with a short summary and open questions of transaction logging as a data collection method.

2.1 INTRODUCTION

The use of transaction logs for research purposes certainly falls conceptually within the confines of the behaviorist paradigm of research. Therefore, behaviorism is the conceptual basis for Web analytics.

2.2 BEHAVORISM

Behaviorism is a research approach that emphasizes the outward behavioral aspects of thought. Strictly speaking, behaviorism also dismisses the inward experiential and procedural aspects [137, 152]; importantly, behaviorism has been heavily criticized for this narrow viewpoint. Some of the pioneers in the behaviorist field are shown in Figure 2.1.

For the area of Web analytics, however, we take a more open view of behaviorism. In this more accepting view, behaviorism emphasizes observed behaviors without discounting the inner aspects (i.e., attitudinal characteristics and context) that may accompany these outward behaviors. This more open outlook of behaviorism supports the position that researchers can gain much from

| Ivan Petrovich Pavlov | John B. Watson | Burrhus Frederic Skinner |

FIGURE 2.1: Three pioneers of behaviorist research. Copyright © 2009 Photo Researchers, Inc. All Rights Reserved. Used with permission.

studying expressions (i.e., behaviors) of users interacting with information systems. These expressed behaviors may reflect aspects of the person's inner self as well as contextual aspects of the environment within which the behavior occurs. These environmental aspects may influence behaviors while also reflecting inner cognitive factors.

The primary proposition underlying behaviorism is that all things that people do are behaviors. These behaviors include utterances, actions, thoughts, and feelings. With this underlying proposition, the behaviorist position is that all theories and models concerning people have observational correlates. Moreover, the behaviors and any proposed theoretical constructs must be mutually complementary.

Strict behaviorism would further state that there are no differences between the publicly observable behavioral processes (i.e., actions) and privately observable behavioral processes (i.e., thinking and feeling). Due to affective, contextual, situational, or environmental factors, however, there may be disconnections between the cognitive and affective processes. Therefore, there are sources of behavior both internal (i.e., cognitive, affective, and expertise) and external (i.e., environmental and situational). Behaviorism focuses primarily on only what an observer can see or manipulate.

Behaviorism is evident in any research where the observable evidence is critical to the research questions or methods, and this is especially true in any experimental research where the "operationalization" of variables is required. A behaviorist approach, at its core, seeks to understand events in terms of behavioral criteria [134, p. 22]. Behaviorist research demands behavioral evidence, and this is particularly important to Web analytics. Within such a perspective, there is no knowable difference between two states unless there is a demonstrable difference in the behavior associated with each state.

Research that is grounded in behaviorism always focuses on *somebody* doing *something* in a *situation*. Therefore, all derived research questions focus on *who* (actors), *what* (behaviors), *when* (temporal), *where* (contexts), and *why* (cognitive). The actors in a behaviorist paradigm are people, at whatever level of aggregation (e.g., individuals, groups, organizations, communities, nationalities, societies), whose behavior is studied. All aspects of what the actors do are studied carefully. These behaviors have a temporal element, and thus researchers need to study when and how long these behaviors occur. Similarly, the behaviors occur within some context, which are all the environmental and situational features in which these behaviors are embedded, and this context must be recognized and analyzed. Finally, the cognitive aspect to these behaviors is the thought and affective processes internal to the actors executing the behaviors.

From this research perspective, each of these aspects (i.e., actor, behaviors, temporal, context, and cognitive) are behaviorist constructs. However, for Web analytics, we are primarily concerned with defining what a behavior is.

2.3 BEHAVIORS

Defining a behavior is not as straightforward as it may seem at first glance, yet defining a behavior is critical for Web analytics. In research, a variable represents a set of events where each event may have a different value. In Web analytics, session duration or number of clicks may be variables that interest a researcher. The particular variables that a researcher is interested in stem from the research questions driving the study.

We can define variables by their use in a research study (e.g., independent, dependent, extraneous, controlled, constant, and confounding) and by their nature. Defined by their nature, there are three types of variables: *environments* (i.e., events of the situation, environment, or context), *subjects* (i.e., events or aspects of the subject being studied), and *behavioral* (i.e., observable events of the subject of interest).

For Web analytics, behavior is the essential construct of the behaviorist paradigm. At its most basic, a behavior is an observable activity of a person, animal, team, organization, or system. Like many basic constructs, behavior is an overloaded term, as it also refers to the aggregate set of responses to both internal and external stimuli. Therefore, behaviors can also address a spectrum of actions. Because of its many associations, it is difficult to characterize a word like behavior without specifying a context in which it takes place to provide the necessary meaning.

However, one can generally classify behaviors into three general categories:

- Behaviors are something that can be detected and, therefore, recorded.
- Behaviors are an action or a specific goal-driven event with some purpose other than the specific action that is observable.
- Behaviors are reactive responses to environmental stimuli.

In some manner, the researcher must observe these behaviors. In other words, the researcher must study and gather information on a behavior concerning what the actor does. Classically, observation is visual, where the researcher uses his/her own eyes, but recording devices, such as a camera, can assist in the observation. Technology has extended the concept of observation to include other recording devices. For Web analytics, we extend the notion of observation to include logging software. Logging software is really nearly invisible to many users; thus, it allows for a more objective measure of true user behavior. Web analytics focuses on descriptive observation and logging the behaviors, as they would occur in a user–system interaction episode.

When studying behavioral patterns with Web analytics and other similar approaches, researchers often use *ethograms*. An ethogram is a taxonomy or index of the behavioral patterns that details the different forms of behavior that a particular user exhibits. In most cases it is desirable to

TABLE 2.1: Taxonomy of user–system behaviors [67].

STATE	DESCRIPTION
View results	Behavior in which the user viewed or scrolled one or more pages from the results listing. If a results page was present and the user did not scroll, we counted this as a View Results Page.
With Scrolling	*User scrolled the results page.*
Without Scrolling	*User did not scroll the results page.*
but No Results in Window	*User was looking for results, but there were no results in the listing.*
Selection	Behavior in which the user makes a selection in the results listing.
Click URL (in results listing)	*Interaction in which the user clicked on a URL of one of the results in the results page.*
Next in Set of Results List	*User moved to the Next results page.*
Previous in Set of Results List	*User moved to the Previous results page.*
GoTo in Set of Results List	*User selected a specific results page.*
View document	Behavior in which the user viewed or scrolled a particular document in the results listings.
With Scrolling	*User scrolled the document.*
Without Scrolling	*User did not scroll the document.*
Execute	Behavior in which the user initiated an action in the interface.
Execute Query	*Behavior in which the user entered, modified, or submitted a query without visibly incorporating assistance from the system. This category includes submitting the original query which was always the first interaction with system.*
Find Feature in Document	*Behavior in which the user used the FIND feature of the browser.*
Create Favorites Folder	*Behavior in which the user created a folder to store relevant URLs.*
Navigation	Behavior in which the user activated a navigation button on the browser, such as Back or Home.

TABLE 2.1: (*continued*)	
STATE	**DESCRIPTION**
Back	*User clicked the Back button.*
Home	*User clicked the Home button.*
Browser	Behavior in which the user opened, closed, or switched browsers.
Open new browser	*User opened a new browser.*
Switch/Close browser window	*User switched between two open browsers or closed a browser window.*
Relevance action	Behavior such as print, save, bookmark, or copy.
Bookmark	*User bookmarked a relevant document.*
Copy–Paste	*User copy–pasted all of, a portion of, or the URL to a relevant document.*
Print	User printed a relevant document.
Save	User saved a relevant document.
View/Implement assistance	Behavior in which the user viewed the assistance offered by the application.
Implement Assistance	*Behavior in which the user entered, modified, or submitted a query, utilizing assistance offered by the application.*
Phrase	*User implemented the PHRASE assistance.*
Spelling	*User implemented the SPELLING assistance.*
Synonyms	*User implemented the SYNONYMS assistance.*
Previous Queries	*User implemented the PREVIOUS QUERIES assistance.*
Relevance Feedback	*User implemented the RELEVANCE FEEDBACK assistance.*
AND	*User implemented the AND assistance.*
OR	*User implemented the OR assistance.*

create an ethogram in which the categories of behavior are objective and discrete, not overlapping with each other. In an ethogram, the definitions of each behavior and category of behaviors should be clear, detailed, and distinguishable from each other. Ethograms can be as specific or general as the study or investigation warrants.

Spink and Jansen [140] and Jansen and Pooch [69] outline some of the key behaviors for SLA, a specific form of Web analytics. Hargittai [52] and Jansen and McNeese [67] present examples of detailed classifications of behaviors during Web searching. As an example, Table 2.1 presents an ethogram of user behaviors interacting with a Web browser during a searching session employed in the study.

There are many way to observe behaviors. In TLA, we are primarily concerned with observing and recording these behaviors in a file, and we then can view the recorded fields as trace data.

2.4 TRACE DATA

In any study, the researcher has several options for collecting data, and there is no one single best method for data collection. The decision about which approach or approaches to use depends upon the research questions (i.e., what needs to be investigated? how one needs to record the data? what resources are available? what is the timeframe available for data collection? how complex is the data? what is the frequency of data collection? and how will the data be analyzed?).

When collecting transaction log data, we are generally concerned with observations of behavior. The general objective of observation is to record the behavior, either in a natural state or in a laboratory study. In both settings, ideally, the researcher should not interfere with the behavior. However, when observing people, the knowledge that they are being observed is likely to alter participants' behavior. For example, in laboratory studies, a researcher's instructions may make a participant either more or less likely to perform a particular behavior, such as smiling or following a Web link. With logging software, the introduction of the application may change a user's behavior. Although in naturalistic settings, the log application has less of an impact if the user does not feel that the data will be used immediately for research purposes.

When investigating user behaviors, the researcher must keep in mind these limitations of observational techniques even while recording behaviors as data for future analysis. The user, a third party, or the researcher can record the behaviors. Transaction logging is an indirect method of recording data about behaviors, and the users themselves, with the help of logging software, make these data records of behavior. We refer to these records as traces. Thus, transaction log records are a source of trace data.

What is trace data? The processes by which people conduct the activities of their daily lives many times create things, leave marks, induce wear, or reduce some existing material. Within the confines of research, these things, marks, and wear become data. Classically, trace data are the physical remains of interaction [154, pp. 35–52]. These remains can be intentional (i.e., notes in a diary or initials on a cave wall) or accidental (i.e., footprints in the mud or wear on a carpet). However, trace data can also be through third party logging applications. In TLA, we are primarily interested in this data from third party logging.

Many researchers use physical or, as in the case of Web analytics, virtual traces as indicators of behavior. These behaviors are the facts or data that researchers use to describe or make inferences about events concerning the actors. Researchers [154] classify trace data into two general types: erosion and accretion. *Erosion* is the wearing away of material leaving a trace. *Accretion* is the buildup of material, making a trace. Both erosion and accretion have several subcategories. In TLA, we are primarily concerned with accretion trace data.

Trace data (a.k.a., trace measures) offer a sharp contrast to data collected directly. The greatest strength of trace data is that it is unobtrusive, meaning the collection of the data does not interfere with the natural flow of behavior and events in the given context. Since the data is not directly collected, there is no observer present where the behaviors occur to affect the participants' actions, and thus, the researcher is getting data that reflects natural behaviors. Trace data is unique; as unobtrusive and nonreactive data, it can make a very valuable research contribution. In the past, trace data was often time consuming to gather and process, making such data costly. With the advent of transaction logging software, trace data for the studying of behaviors of users and systems in Web analytics is much cheaper to collect, and consequently, Web analytics and related fields of study have really taken off.

Interestingly, in the physical world, erosion data is what typically reveals usage patterns (i.e., trails worn in the woods, footprints in the snow, fingerprints on a book cover). However, with Web analytics, logged accretion data indicates the usage patterns (i.e., access to a Website, submission of queries, Webpages viewed). Specifically, transaction logs are a form of controlled accretion data, where the researcher or some other entity alters the environment in order to create the accretion data [154, pp. 35–52]. With a variety of tracking applications, the Web is a natural environment for controlled accretion data collection. With the user of client apps (such as desktop search bars and what not), the collection of data is nearly unlimited from a technology perspective.

Like all data collection methods, trace data for studying users and systems has strengths and limitations. Certainly, trace data are valuable for understanding behavior (i.e., behavioral actions) in naturalistic environments and may offer insights into human activity obtainable in no other way. For example, data from Web transaction logs is on a scale available in few other places. However, one must interpret trace data carefully and with a fair amount of caution because trace data can be incomplete or even misleading. For example, with the data in transaction logs the researcher can say a given number of Website users only looked at the Website's homepage and then left (a.k.a., homepage bounce rate). However, using trace data alone the researcher could not conclude whether the users left because they found what they were looking for, were frustrated because they could not find what they were looking for, or were in the wrong place to begin with. However, with some experimental data one could make some reasonable assumptions concerning this user behavior.

Research using trace data from transaction logs should be analyzed based on the same criteria as all research data and methods. These criteria are *credibility*, *validity*, and *reliability*.

Credibility concerns how trustworthy or believable the data collection method is. The researcher must make the case that the data collection approach records the data needed to address the underlying research questions.

Validity addresses whether the measurement actually measures what it is supposed to measure. There are three kinds of validity:

- **Face or internal validity**: the extent to which the contents of the test, method, analysis, or procedure that the researcher is employing measure what they are supposed to measure.
- **Content or construct validity**: the extent to which the content of the test, method, analysis, or procedure adequately represents all that is required for validity of the test, method, analysis, or procedure (i.e., are you collecting and accounting for all that you should collect and account for).
- **External validity**: the extent to which one can generalize the research results across populations, situations, environments, and contexts of the test, method, analysis, or procedure.

In inferential or predictive research, one must also be concerned with statistical validity (i.e., the degree of strength of the independent and dependent variable relationships). Statistical validity is actually an important aspect for Web analytics, given the needed ties between data collected and KPIs.

Reliability is a term used to describe the stability of the measurement. Essentially, reliability addresses whether the measurement assesses the same thing, in the same way, in repeated tests.

Researchers must always address the issues of credibility, validity, and reliability. Leveraging the work of Holst [58], the researcher must address six questions in every Web analytics research project that uses trace data from transaction logs.

- **Which data are analyzed?** The researcher must clearly communicate in a precise manner both the format and content of recorded trace data. With transaction log software, this is much easier than in other forms of trace data, as logging applications can be reverse engineered to articulate exactly what behavioral data is recorded.
- **How is this data defined?** The researcher must clearly define each trace measure in a manner that permits replication of the research on other systems and with other users. As TLA

has proliferated in a variety of venues, more precise definitions of measures are developing [114, 151, 158].

- **What is the population from which the researcher has drawn the data?** The researcher must be cognizant of the actors, both people and systems, that created the trace data. With transaction logs on the Web, this is sometimes a difficult issue to address directly, unless the system requires some type of logon and these profiles are then available. In the absence of these profiles, the researcher must rely on demographic surveys, studies of the system's user population, or general Web demographics.

- **What is the context in which the researcher analyzed the data?** It is important for the researcher to explain clearly the environmental, situational, and contextual factors under which the trace data was recorded. With transaction log data, this includes providing complete information about the temporal factors of the data collection (i.e., the date and time the data was recorded) and the make-up of the system at the time of the data recording, as system features undergo continual change. Transaction logs have the significant advantage of time sampling of trace data. In time sampling, the researcher can make the observations at predefined points of time (e.g., every 5 minutes, every second), and then record the action that is taking place, using the classification of action defined in the ethogram.

- **What are the boundaries of the analysis?** Research using trace data from transaction logs is tricky, and the researcher must be careful not to overreach with the research questions and findings. The implications of the research are confined by the data and the method of the data collected. For example, with transaction log data we can rather clearly state whether or not a user clicked on a link. However, transaction log trace data itself will not inform us as to why the user clicked on a link. Was it intentional? Was it a mistake? Did the user become sidetracked?

- **What is the target of the inferences?** The researcher must clearly articulate the relationship among the separate measures in the trace data either to inform descriptively or in order to make inferences. Trace data can be used for both descriptive research to improve our understanding and predictive research in terms of making inferences. These descriptions and inferences can be at any level of granularity (i.e., individual, collection of individuals, organization, etc.). However, Hilbert and Redmiles [55] point out, based on their experiences, that transaction log data is best used for aggregate level analysis. I disagree with this position. With enough data at the individual level, one can tell a lot from log data.

If the researcher addresses each of the six questions, transaction logs are an excellent way to collect trace data on users of Web and other information systems. The researcher then examines this data using TLA. The use of trace data to understand behaviors makes the use of transaction logs and transaction logs analysis an unobtrusive research method.

2.5   UNOBTRUSIVE METHODS

As noted in the introduction to this lecture, unobtrusive methods are research practices that do not require the researcher to intrude in the context of the actors and thus do not involve direct elicitation of data from the research participants or actors. Unobtrusive measurement presumably reduces the biases that result from the intrusion of the researcher or measurement instrument. We should note, however, that unobtrusive measures reduce the degree of control that the researcher has over the type of data collected, and importantly, for some research questions, appropriate unobtrusive measures may simply not be available.

Why is it important for the research not to intrude upon the environment? There are at least three justifications. First, the Heisenberg uncertainty principle, borrowed from the field of quantum physics asserts that the outcome of a measurement of some system is neither deterministic nor perfect. Instead, a measurement is characterized by a probability distribution. The larger the associated standard deviation is for this distribution, the more "uncertain" are the characteristics measured for the system. The Heisenberg uncertainty principle is commonly stated as, "One cannot accurately and simultaneously measure both the position and momentum of a mass" (http://en.wikipedia .org/wiki/Uncertainty_principle). In this analogy, when researchers are interjected into an environment, they become part of the system. Therefore, their very presence in the environment will affect measurements of the components of that system. A common example is in ethnographic studies where the researchers interject themselves in a given context.

The second justification for avoiding or at least limiting environmental intrusion is the observer effect. The observer effect refers to the difference that is made to an activity or a person's behaviors by it (or the person) being observed. People may not behave in their usual manner if they know that they are being watched or when being interviewed while carrying out an activity. In research, this observer effect specifically refers to changes that the act of observing will make on the phenomenon being observed. In information technology, the observer effect is the potential impact of the act of observing a process output while the process is running. A good example of the observer effect in TLA is pornographic searching behavior. Participants rarely search for porn in a laboratory study while studies employing trace data shows it is a common searching topic [71].

In addition to the uncertainty principle and the observer effect, observer bias adds a third justification for reducing environmental intrusion. Observer bias is error that the researcher introduces into measurement when observers overemphasize behavior they expect to find and fail to notice behavior they do not expect. Many fields have common procedures to address this, although these procedures are seldom used in information and computer science. For example, the observer bias is why medical trials are normally double-blind rather than single-blind. Observer bias is introduced because researchers see a behavior and interpret it according to what it means to them, whereas it may mean something else to the person showing the behavior. Trace data helps in overcoming the observer bias in the data collection. However, as with other methods, trace data has no effect on the observer bias in interpreting the results from data analysis.

Given the justifications for using unobtrusive methods, we will now turn our attention to three types of unobtrusive measurement that are applicable to Web analytics, namely indirect analysis, context analysis, and second analysis. Web analytics is an indirect analysis method. The researcher is able to collect the data without introducing any formal measurement procedure. In this regard, TLA typically focuses on the interaction behaviors occurring among the users, system, and information. There are several examples of utilizing transaction analysis as an indirect approach [cf. Refs. 2, 15, 32, 57].

Content analysis is the analysis of text documents. The analysis can be quantitative, qualitative, or a mixed methods approach. Typically, the major purpose of content analysis is to identify patterns in text. Content analysis has the advantage of being unobtrusive and, depending on whether automated methods exist, can be a relatively rapid method for analyzing large amounts of text. In Web analytics, content analysis typically focuses on search queries or analysis of retrieved results. A variety of examples are available in this area of transaction log research [cf. Refs. 7, 16, 51, 151, 158].

Secondary data analysis, like content analysis, makes use of already existing sources of data. However, secondary analysis typically refers to the re-analysis of quantitative data rather than text. Secondary data analysis uses data that was collected by others to address different research questions or to use different methods of analysis than was originally intended during data collection. For example, Websites commonly collect transaction log data for system performance analysis. However, researchers can also use this data to address other questions. Several transaction log studies have focused on this aspect of research [21, 22, 29, 30, 34, 77, 107, 129].

As a secondary analysis method, Web analytics has several advantages. First, it is efficient in that it makes use of data collected by a Website application. Second, it often allows the researcher to extend the scope of the study considerably by providing access to a potentially large sample of users over a significant duration [81]. Third, since the data is already collected, the cost of using existing transaction log data is cheaper than collecting primary data.

However, the use of secondary analysis is not without difficulties. First, secondary data is frequently not trivial to prepare, clean, and analyze [66], especially large transaction logs. Second, researchers must often make assumptions about how the data was collected because third parties developed the logging applications. A third and perhaps more perplexing difficulty concerns the ethics of using transaction logs as secondary data. By definition, the researcher is using the data in a manner that may violate the privacy of the system users [53]. In fact, some critics point to a growing concern for unobtrusive methods due to increased sensitivity toward the ethics involved in such research [112]. Log data may be unobtrusive, but it can certainly be quite invasive.

2.6 WEB ANALYTICS AS UNOBTRUSIVE METHOD

Web analytics has significant advantages as a methodological approach for the study and investigation of behaviors. These factors include:

- **Scale**: Transaction log applications can collect data to a degree that overcomes the critical limiting factor in laboratory user studies. User studies in laboratories are typically restricted in terms of sample size, location, scope, and duration.
- **Power**: The sample size of transaction log data can be quite large, so inference testing can highlight statistically significant relationships. Interestingly, sometimes the amount of data in transaction logs from the Web is so large that nearly every relation is significantly correlated. Due to the large power, researchers must account for the size effect.
- **Scope**: Since transaction log data is collected in natural contexts, researchers can investigate the entire range of user–system interactions or system functionality in a multi-variable context.
- **Location**: Transaction log data can be collected in naturalistic, distributed environments. Therefore, users do not have to be in an artificial laboratory setting.
- **Duration**: Since there is no need for recruiting specific participants for a user study, transaction log data can be collected over an extended period.

All methods of data collection have strengths not available with other methods, but they also have inherent limitations. Transactions logs have several shortcomings. First, transaction log data is not nearly as versatile relative to primary data because the data may not have been collected with the particular research questions in mind. Second, transaction log data is not as rich as some other data collection methods and therefore not available for investigating the range of concepts some researchers may want to study. Third, the fields that the transaction log application records are many times only loosely linked to the concepts they are alleged to measure. Fourth, with transaction logs the users may be aware that they are being recorded and may alter their actions. Therefore, the user behaviors may not be altogether natural.

Given the inherent limitations in the method of data collection, Web analytics also suffers from shortcomings derived from the characteristics of the data collection. Hilbert and Redmiles [56] maintain that all research methods suffer from some combination of abstraction, selection, reduction, context, and evolution problems that limit scalability and quality of results. Web analytics suffers from these same five shortcomings.

- **Abstraction problem**—how does one relate low-level data to higher-level concepts?
- **Selection problem**—how does one separate the necessary from unnecessary data before reporting and analysis?

- **Reduction problem**—how does one reduce the complexity and size of the data set before reporting and analysis?
- **Context problem**—how does one interpret the significance of events or states within state chains?
- **Evolution problem**—how can one alter data collection applications without impacting application deployment or use?

Because each method has its own combination of abstraction, selection, reduction, context, and evolution problems, astute researchers will employ complementary methods of data collection and analysis. This is similar to the conflict inherent in any overall research approach. Each research method for data collection tries to maximize three desirable criteria: *generalizability* (i.e., the degree to which the data applies to overall populations), *precision* (i.e., the degree of granularity of the measurement), and *realism* (i.e., the relation between the context in which evidence is gathered relative to the contexts to which the evidence is to be applied). Although the researcher always wants to maximize all three of these criteria simultaneously, in reality it cannot be done. This is one fundamental dilemma of the research process. The very things that increase one of these three features will reduce one or both of the others.

2.7   CONCLUSION

Recordings of behaviors via transaction log applications on the Web opens a new era for researchers by making large amounts of trace data available for use. The online behaviors and interactions among users, systems, and information create digital traces that permit collection and analysis of this data. Logging applications provide data obtained through unobtrusive methods, and importantly, these collections are substantially larger than any data set obtained via surveys or laboratory studies. As noted earlier, these applications allow the data to be collected in naturalistic settings with little to no impact by the observer. Researchers can use these digital traces to analyze a nearly endless array of behavior topics.

Web analytics is a behaviorist research method, with a natural reliance on the expressions of interactions as behaviors. The transaction log application records these interactions, creating a type of trace data. As a reminder, trace data in transaction logs are records of interactions as people use these systems to locate information, navigate Websites, and execute services. The data in transaction logs is a record of user–system, user–information, or system–information interactions. Moreover, transaction logs provide an unobtrusive method of collecting data on a scale well beyond what one could collect in confined laboratory studies. Figure 2.2 provides a recap of the foundation of Web analytics.

The massive increased availability of Web trace data has sparked concern over the ethical aspects of using unobtrusively obtained data from transaction logs. For example, who does the

Type of Data	Trace	
Data Collection	Unobtrusive	
Key Construct	Behavior	
Theoretical Foundation	Behaviorism	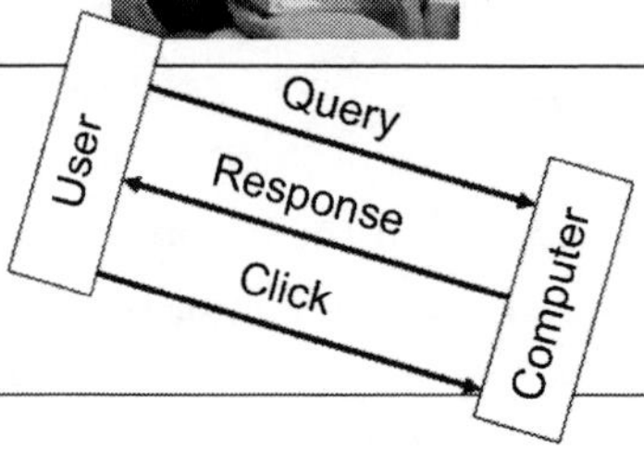

FIGURE 2.2: Recap of foundational element of Web analytics.

trace data belong to—the user, the Website that logged the data, or the public domain? How does (or should) one seek consent to use such data? If researchers do seek consent, from whom does the researcher seek it? Is it realistic to require informed consent for unobtrusively collected data? These are open questions.

*   *   *   *

CHAPTER 3

The History of Web Analytics

There have been an increasing number of review articles on Web analytics research in academia. One of the first, Jansen and Pooch [69] provide a review of Web transaction log research of Web search engines and individual Websites through 2000, focusing on query analysis. After reviewing studies conducted between 1995 and 2000, Hsieh-Yee [59] reports that many studies investigate the effects of certain factors on Web search behavior, including information organization and presentation, type of search task, Web experience, cognitive abilities, and affective states. Hsieh-Yee [59] also notes that many studies lack external validity.

Bar-Ilan [13] presents an extensive and integrative overview of Web search engines and the use of Web search engines in information science research. Bar-Ilan [13] provides a variety of perspectives including user studies, social aspects, Web structure, and search-engine evaluation.

Two excellent historical reviews are Penniman [115, 116], who examines log research from the very beginning as a participant/observer, and Markey [98, 99], who reviews twenty-five years of academic research in the area.

Given the availability of these comprehensive reviews, we will touch on some of the previous work simply to identify the overall trends and to provide historical insight for Web analytics today. Web analytics studies fall into three categories: (1) those that primarily use transaction-log analysis, (2) those that incorporate users in a laboratory survey or other experimental setting, and (3) those that examine issues related to or affecting Web searching.

3.1 SINGLE WEBSITES

Some researchers have used transaction logs to explore user behaviors on single Websites. For example, Yu and Apps [162] used transaction log data to examine user behavior in the Super-Journal project. For 23 months (February 1997 to December 1998), the researchers recorded 102,966 logged actions, related these actions to 4 subject clusters, 49 journals, 838 journal issues, 15,786 articles, and 3 Web search engines.

In another study covering the period from 1 January to 18 September 2000, Kea et al. [82] examined user behavior in Elsevier's ScienceDirect, which hosts bibliographic information and full-text articles of more than 1300 journals with an estimated 625,000 users. Loken et al. [96] examined the transaction log data of the online self-directed studying of more than 100,000 students using a

Web-based system to prepare for U.S. college admissions tests. The researchers noted several non-optimal behaviors, including a tendency toward deferring study and a preference for short-answer verbal questions. The researchers discussed the relevance of their findings for online learning.

Wen et al. [156] investigated the use of click-through data to cluster queries for question answering on a Web-based version of the Encarta encyclopedia. The researchers explored the similarity between two queries using the common user-selected documents between them. The results indicate that a combination of both keywords and user logs is better than using either method alone. Using a Lucent proxy server, Hansen and Shriver [50] used transaction-log analysis to cluster search sessions and to identify highly relevant Web documents for each query cluster.

Collectively, these studies provide better descriptions of user behaviors and help to refine transaction log research for Web log analysis of searching and single Websites from an academic perspective.

3.2 LIBRARY SYSTEMS

Some of the original work in the area of log analysis has occurred in the library fields, with many studies of library systems [117]. Continuing this rich tradition of using transaction logs to investigate the use of library systems, more sophisticated methods are emerging. For example, Chen and Cooper [28] clustered users of an online library system into groups based on patterns of states using transaction logs data. The researchers defined 47 variables, using them to classify 257,000 sessions. Then they collapsed these 47 variables into higher order groupings, identifying six distinct clusters of users. In a follow-up study, Chen and Cooper [27] used 126,925 sessions from the same online system, modeling patterns using Markov models. The researchers found that a third-order Markov model explained five of the six clusters.

What is coming out of this line of research is a move in the academic field from straightly descriptive to more predictive elements of analysis, including methods of Web mining [136].

3.3 SEARCH ENGINES

Rather than focusing on single Websites, other researchers have investigated information searching on Web search engines. Ross and Wolfram [131] analyzed queries submitted to the Excite search engine for subject content based on the co-occurrence of terms. The researchers categorized more than 1000 of the most frequently co-occurring term pairs into one or more of 30 developed subject areas. The cluster analyses resulted in several well-defined high-level clusters of broad subject areas. He et al. [54] examined contextual information from Excite and Reuters transaction logs, using a version of the Dempster–Shafer theory to identify search engine sessions. The researchers determined the average Web user session duration was about 12 min. Özmutlu and Cavdur [109] investigated contextual information using an Excite transaction log. The researchers explored the reasons underlying the inconsistent performance of automatic topic identification with statistical analysis and experimental design techniques. Xie and O'Hallaron [160] investigated caching to reduce both server load and user-response time in distributed systems by analyzing a transaction log from the Vivisimo search engine, from 14 January to 17 February 2001. The researchers report that queries have significant locality, with query frequency following a Zipf distribution. Lempel and Moran [92] also investigated clustering to improve caching of search engine results using more than seven million queries submitted to AltaVista. The researchers report that pre-fetching of search engine results can increase cache–hit ratios by 50 percent for large caches and can double the hit ratios of small caches. There is much ongoing work in the area of using logs for search engine and server caching [10].

In what appears currently to be one of the longest temporal studies, Wang et al. [151] analyzed 541,920 user queries submitted to an academic-Website search engine during a four-year period (May 1997 to May 2001). Conducting analysis at the query and term levels, the researchers report that 38% of all queries contained only one term and that most queries are unique. Eiron and McCurley [38] used 448,460 distinct queries from an IBM Intranet search engine to analyze the effectiveness of anchor text.

Pu [122] explored the searching behavior of users searching on two Taiwanese Web search engines, Dreamer and Global Area Information Servers (GAIS). The average length of English terms on these two Web search engines was 1.0 term for Dreamer and 1.22 terms for GAIS. Baeza-Yates and Castillo [9] examined approximately 730,000 queries from TodoCL, a Chilean search system. They found that queries had an average length of 2.43 terms. A lengthier analysis is presented in Baeza-Yates and Castillo [8]. Montgomery and Faloutsos [107] analyzed more than 20,000 Internet users who accessed the Web from July 1997 through December 1999 using data provided by Jupiter Media Metrix (http://www.jupiterresearch.com). The researchers report users revisited 54 percent of URLs at least once during a searching session.

They also report that browsing patterns follow a power law and the patterns remained stable throughout the period of analysis. Rieh and Xu [127] analyzed queries from 1,451,033 users of Excite collected on 9 October 2000. The researchers examined how each user reformulated his/ her Web query over a 24-hour period. Out of the 1,451,033 users logs collected, the researcher used various criteria to select 183 sessions for manual analysis. The results show that while most query reformulation involves content changes, about 15% of the reformulation relate to format modifications.

Huang et al. [60] propose an effective term-suggestion approach for interactive Web search using more than two million queries submitted to Web search engines in Taiwan. The researchers propose a transaction log approach to relevant term extraction and term suggestion using relevant terms that co-occur in similar query sessions.

Jansen and Spink [70] determined that the typical Web searching session was about 15 min from an analysis of click through data from AlltheWeb.com. The researchers report that the Web search engine users on average view about eight Web documents, with more than 66% of searchers examining fewer than five documents in a given session. Users on average view about two to three documents per query. More than 55% of Web users view only one result per query. Twenty percent of the Web users view a Web document for less than a minute. These results would seem to indicate that the initial impression of a Web document is extremely important to the user's perception of relevance.

Beitzel et al. [15] examine hundreds of millions of queries submitted by approximately 50 million users to America Online (AOL) over a 7-day period from 26 December 2003 through 1 January 2004. During this period, AOL used results provided by Google. The researchers report that only about 2% of the queries contain query operators. The average query length is 2.2 terms, and 81% of users view only one results page. The researchers report changes in popularity and uniqueness of topically categorized queries across hours of the day. Park, Bae, and Lee (Forthcoming) analyzed transaction logs of NAVER, a Korean Web search engine and directory service. The data was collected over a one-week period, from 5 January to 11 January 2003, and contained 22,562,531 sessions and 40,746,173 queries. Users of NAVER implement queries with few query terms, seldom use advanced features, and view few results pages. Users of NAVER had an average session length of 1.8 queries. Wolfram et al. [159] analyze session clusters from three different search environments.

Web analytics is also entering a variety of areas, including keyword advertising and sponsored search [65].

Clearly, these research projects provide valuable information for understanding and perhaps improving user–system and system–information interactions.

3.4   CONCLUSION

What does a historical review of transactional log analysis inform us about the current and possible future state of Web analytics? In one of the earliest studies employing transaction logs, Penniman [116, p. 159] stated, "The promise (of transaction logs) is unlimited for evaluating communicative behavior where human and computer interact to exchange information." Since the mid-1960s, we have seen the use of transaction logs evolve from an almost purely descriptive approach focusing primarily on system effectiveness to one focusing on the combined aspects of the both user and system. Today, we see these tools being leveraged for more insightful and predictive aspects of the user–system interaction. Combined with associated research methods, transaction logs have served a vital function in understanding users and systems.

· · · ·

CHAPTER 4

Data Collection for Web Analytics

As the previous brief review of research demonstrates, data for Web analytics is plentiful. How the data is collected, however, is important. There is a proliferation of techniques (e.g., performance monitors, Web server log files, cookies, and packet sniffing), but the most common individual techniques generally fall into one of two major approaches for collecting data for Web analysis: log files and page tagging [80]. Most current Web analytic companies use a combination of the two methods for collecting data. Therefore, anyone interested in Web analytics needs to understand the strengths and weaknesses of each.

4.1 WEB SERVER LOG FILES

The first method of metric gathering uses log files. Every Web server keeps a log of page requests that can include (but is not limited to) visitor Internet Protocol (IP) address, date and time of the request, request page, referrer, and information on the visitor's Web browser and operating system. The same basic collected information can be displayed in a variety of ways. Although the format of the log file is ultimately the decision of the company who runs the Web server, the following four formats are a few of the most popular: *NCSA Common Log*, *NCSA Combined Log*, *NCSA Separate Log*, and *W3C Extended Log*.

The NCSA Common Log format (also known as Access Log format) contains only basic information on the page request. This includes the client IP address, client identifier, visitor username, date and time, HTTP request, status code for the request, and the number of bytes transferred during the request. The Combined Log format contains the same information as the common log with the following three additional fields: the referring URL, the visitor's Web browser and operating system information, and the cookie. The Separate Log format (or 3-Log format) contains the same information as the combined log, but it breaks it into three separate files—the access log, the referral log, and the agent log. The date and time fields in each of the three logs are the same. Table 4.1 shows examples of the common, combined, and separate log file formats (notice that default values are represented by a dash (-).

Similarly, W3C provides an outline for standard formatting procedures. This format differs from the first three in that it aims to provide for better control and manipulation of data while

TABLE 4.1: NCSA log comparison.

TYPE OF LOG	EXAMPLE ENTRY
NCSA Common Log	111.222.125.125 - jimjansen [10/Oct/2009:21:15:05 +0500] "GET /index.html HTTP/1.0" 200 1043
NCSA Combined Log	111.222.125.125 - jimjansen [10/Oct/2009:21:15:05 +0500] "GET /index.html HTTP/1.0" 200 1043 "http://ist.psu.edu/faculty_pages/ jjansen/" "Mozilla/4.05 [en] (WinNT; I)" "USERID=CustomerA; IMPID=01234"
NCSA Separate Log	Common Log: 111.222.125.125 - jimjansen [10/Oct/2009:21:15:05 +0500] "GET /index.html HTTP/1.0" 200 1043 Referral Log: [10/Oct/2009:21:15:05 +0500] "http://ist.psu.edu/faculty_pages/ jjansen/" Agent Log: [10/Oct/2009:21:15:05 +0500] "Microsoft Internet Explorer - 7.0"

TABLE 4.2: W3C extended log file.

TYPE OF LOG	EXAMPLE ENTRY
W3C Extended Log	#Software: Microsoft Internet Information Services 6.0 #Version: 1.0 #Date: 2009 -05-24 20:18:01 #Fields: date time c-ip cs-username s-ip s-port cs-method cs-uri-stem cs-uri-query sc-status sc-bytes cs-bytes time-taken cs(User-Agent) cs(Referrer) 2009-05-24 20:18:01 172.224.24.114 - 206.73.118.24 80 GET /Default.htm - 200 7930 248 31 Mozilla/ 4.0+(compatible;+MSIE+7.01;+Windows+2000+Server) http://54.114.24.224/

still producing a log file readable by most Web analytics tools. The extended format contains user defined fields and identifiers followed by the actual entries, and default values are represented by a dash (-). Table 4.2 shows an example of an extended log file.

System log files offer several benefits for gathering data for analysis. First, using system log files does not require any changes to the Website or any extra software installation to create the log files. Second, because Web servers automatically create these logs and store them on a company's own servers, the company has freedom to change their Web analytics tools and strategies at will. Additionally, using system log files does not require any extra bandwidth when loading a page, and since everything is recorded server-side, it is possible to log both page request successes and failures.

Using log files also has some disadvantages. One major disadvantage is that the collected data is limited to only transactions with the Web server. This means that they cannot log information independent from the servers, such as the physical location of the visitor. Similarly, while it is possible to log cookies, the server must be specifically configured to assign cookies to visitors in order to do so. The final disadvantage is that while it is useful to have all the information stored on a company's own servers, the log file method is only available to those who own their Web servers.

4.2   PAGE TAGGING

The second method for recording data for Web analytics is page tagging. In page tagging, an invisible image is used to detect when a page has been successfully loaded and then triggers JavaScript to send information about the page and the user back to a remote server. According to Peterson [118] the variables used and amount of data collected in page tagging are dependent on the Web analytics vendor. Some vendors stress short, easy to use page tags while others emphasize specific tags that require little post-processing. The best thing to look for with this method, however, is flexibility—being able to use all, part, or none of the tag depending on the needs of the page.

This method of gathering user data offers several benefits. The first is speed of reporting. Unlike a log file, the data received via page tagging is parsed as it comes in. This allows for near real-time reporting. Another benefit is flexibility of data collection. More specifically, it is easier to record additional information about the user without involving a request to the Web server. Examples of such information include information about a user's screen size, the price of purchased goods, and interactions within Flash animations. This is also a useful method of gathering data for companies that do not run their own Web servers or do not have access to the raw log files for their site (such as blogs).

Page tagging also entails some disadvantages, most of which are centered on the extra code that must be added to the Website. This extra code requires the page to use more bandwidth each time it loads, and it also makes it harder to change analytics tools because the code embedded in the Website would have to be changed or deleted entirely. The final disadvantage is that page tagging is

TABLE 4.3: Web server log files versus page tagging [19].

LOG FILES (ALL TYPES)		PAGE TAGGING (ALL APPROACHES)	
ADVANTAGES	*DISADVANTAGES*	*ADVANTAGES*	*DISADVANTAGES*
Does not require changes to the Website or extra hardware installation	Can only record interactions with the Web server	Near real-time reporting	Requires extra code added to the Website
Does not require extra bandwidth	Server must be configured to assign cookies to visitors	Easier to record additional information	Uses extra bandwidth each time the page loads
Freedom to change tools with a relatively small amount of hassle	Only available to companies who run their own Web servers	Able to capture visitor interactions within Flash animations	Can only record successful page loads, not failures
Logs both page request successes and failures	Cannot log physical location		Hard to switch analytic tools

only capable of recording page loads, not page failures. If a page fails to load, it means that the tagging code also did not load, and there is therefore no way to retrieve information in that instance.

Although log files and page tagging are two distinct ways to collect information about the Website users or visitors, it is possible to use both together, and many analytics companies provide ways to use both methods to gather data. Even so, it is important to understand the strengths and weaknesses of both. Table 4.3 presents the advantages and disadvantages of log file analysis and page tagging.

4.3   CONCLUSION

Regardless of whether log files or page tagging is used (or new approaches that may be developed), the data will eventually end up in a log file for analysis. In other words, while the data collection may differ, the method of analysis remains the same.

CHAPTER 5

Web Analytics Fundamentals

To understand and derive the benefits of Web analysis, one must first understand metrics, the different kinds of measures available for analyzing user information [19, 111]. Although metrics may seem basic, once collected we can use these metrics to analyze Web traffic and improve a Website to meet better the expectations of the site's traffic. These metrics generally fall into one of four categories: site usage, referrers (or how visitors arrived at the site), site content analysis, and quality assurance. Table 5.1 shows examples of types of metrics that we might find in these categories.

Although the type and overall number of metrics varies with different analytics vendors, a set of basic metrics is common to most. Table 5.2 outlines eight widespread types of information [63] that measure who is visiting a Website and what they do during their visits, relating each of these metrics to specific categories.

Each metric is discussed below.

5.1   VISITOR TYPE

Since analyzing Website traffic first became popular in the 1990s with the Website counter, the measure of Website traffic has been one of the most closely watched metrics. This metric, however, has evolved from merely counting the number of hits a page receives into counting the number of individuals who visit the Website.

Ignoring the software robots that can make up a large portion of traffic [68], there are two types of visitors: new visitors, meaning those who have not previously visited the site, and repeat visitors, meaning those who have been to the site previously. In order to track visitors in such a way, a system must be able to determine individual users who access a Website; each individual user is called a unique visitor. Ideally, a unique visitor is just one visitor, but this is not always the case. It is possible that multiple users access the site from the same computer (perhaps on a shared household computer or a public library). In addition, most analytic software relies on cookies to track unique users. If a user disables cookies in the browser or if they clear their cache, then the visitor will be counted as new each time he or she enters the site.

Because of this, some companies have instead begun to track unique visits or sessions. A session begins once a user enters the site and ends when a user exits the site or after a set amount of time of inactivity (usually 30 minutes). The session data does not rely on cookies and can be

TABLE 5.1: Metrics categories [63].

SITE USAGE	REFERRERS	SITE CONTENT ANALYSIS	QUALITY ASSURANCE
• Geographic information • How many people repeatedly visit the site • Numbers of visitors and sessions • Search engine activity	• How many people place bookmarks to the site • The search terms people used to find your site • Which Websites are sending visitors to your site	• Effectiveness of key content • Most popular pages • Top entry pages • Top exit pages • Top pages for single page view sessions • Top paths through the site	• Broken pages or server errors • Visitor response to errors

TABLE 5.2: Eight common metrics of Website analysis [19].

METRIC	DESCRIPTION	CATEGORY
Demographics and System Statistics	The physical location and information of the system used to access the Website	Site Usage
Errors	Any errors that occurred while attempting to retrieve the page	Quality Assurance
Internal Search Information	Information on keywords and results pages viewed using a search engine embedded in the Website	Site Usage
Referrering URL and Keyword Analysis	Which sites have directed traffic to the Website and which keywords visitors are using to find the Website	Referrers
Top Pages	The pages that receive the most traffic	Site Content Analysis
Visit Length	The total amount of time a visitor spends on the Website	Site Usage
Visitor Path	The route a visitor uses to navigate through the Website	Site Content Analysis
Visitor Type	Who is accessing the Website (returning, unique, etc.)	Site Usage

measured easily. Since there is less uncertainty with visits, it is considered to be a more concrete and reliable metric than unique visitors. This approach is also more sales-oriented because it considers each visit an opportunity to convert a visitor into a customer instead of looking at overall customer behavior [17].

5.2 VISIT LENGTH

Also referred to as Visit Duration or Average Time on Site, visit length is the total amount of time a visitor spends on a site during one session. One possible area of confusion when using this metric is handling missing data. This can be caused either by an error in data collection or by a session containing only one page visit or interaction. Since the visit length is calculated by subtracting the time of the visitor's first activity on the site from the time of the visitor's final activity, when one of those pieces of data is missing, according to the WAA, the visit length is calculated as zero [23].

When analyzing the visit length, the measurements are often broken down into chunks of time. StatCounter, for example, uses the following time categories [63]:

- Less than 5 seconds
- 5 seconds to 30 seconds
- 30 seconds to 5 minutes
- 5 minutes to 20 minutes
- 20 minutes to 1 hour
- Greater than 1 hour

The goal of measuring the data in this way is to keep the percentage of visitors who stay on the Website for less than five seconds as low as possible. If visitors stay on a Website for such a short amount of time, either they arrived at the site by accident or the site did not have relevant information. By combining this information with information from referrers and keyword analysis, we can determine which sites are referring well-targeted traffic and which sites are referring poor quality traffic.

5.3 DEMOGRAPHIC AND SYSTEM STATISTICS

For some companies, well-targeted traffic means region-specific traffic. For example, if an e-commerce site can only ship its goods to people in Spain, any traffic to the site from outside of Spain is irrelevant. The demographic metric refers to the physical location of the system used to make a page request. This information can be useful for a Website that provides region-specific services. In addition, region-specific Websites also want to make sure they tailor their content to the group they are targeting; thus, demographic information can also be combined with information on referrers to determine if a referral site is directing traffic to a site from outside a company's regions of service.

In addition to demographic location, companies also need information about the hardware and software with which visitors access a Website, and system statistics provide information such as browser type, screen resolution, and operating system. By using this information, companies can tailor their Websites to meet visitors' technical needs, thereby ensuring that all customers can access the Websites.

5.4   INTERNAL SEARCH

If a Website includes a site-specific search utility, then it is also possible to measure internal search information. This can include not only keywords but also information about which results pages visitors found useful. There are several uses for analyzing internal search data [3]:

- Identify products and services for which customers are looking, but that are not yet provided by the company
- Identify products that are offered, but which customers have a hard time finding
- Identify customer trends
- Improve personalized messages by using the customers' own words
- Identify emerging customer service issues
- Determine if customers are provided with enough information to reach their goals
- Make personalized offers

By analyzing internal search data, we can use the information to improve and personalize the visitors' experiences.

5.5   VISITOR PATH

Excluding visitors who leave the site as soon as they enter, each visitor creates a path of page views and actions while perusing the site. By studying these paths, we can identify any difficulties a user has viewing a specific area of the site or completing a certain action (such as making a transaction or completing a form).

According to an article by the WAA, there are two schools of thought regarding visitor path analysis. The first is that visitor actions are goal-driven and performed in a logical, linear fashion. For example, if a visitor wants to purchase an item, the visitor will first find the item, add it to the cart, and proceed to the checkout to complete the process. Any break in that path (i.e., not completing the order) signifies user confusion and is viewed as a problem.

The second school of thought is that visitor actions are random and illogical and that the only path that can provide accurate data on a visitor's behavior is the path from one page to the page immediately following it. In other words, the only page that influences visitors' behavior on a Website is the one they are currently viewing. For example, visitors on a news site may merely peruse the articles with no particular goal in mind. This method of analysis is becoming increasingly popular because companies find it easier to examine path data in context without having to reference the entire site in order to study the visitors' behavior.

5.6   TOP PAGES

The first page a visitor views makes the greatest impression about the Website. These first pages are called top pages and generally fall into three categories: *top entry pages*, *top exit pages*, and *most popular pages*. By knowing the top entry page, organizations can ensure that the page has relevant information and provides adequate navigation to important parts of the site. Similarly, identifying popular exit pages makes it easier to pinpoint areas of confusion or missing content.

The most popular pages are the areas of a Website that receive the most traffic. This metric gives insight into how visitors are utilizing the Website and which pages are providing the most useful information. This kind of information shows whether the Website's functionality matches up with the organization's goals; if most of the Website's traffic is being directed away from the main pages of the site, the Website cannot function to its full potential.

5.7   REFERRERS AND KEYWORD ANALYSIS

Users often reach Websites through a referral page, the page visited immediately before entering a Website, or rather, a site that has directed the user (a.k.a., traffic) to the Website. A search engine result page link, a blog entry mentioning the Website, and a personal bookmark are examples of referrers. By using this metric, organizations can determine advertising effectiveness and search engine popularity. As always, it is important to look at this information in context. If a certain referrer is doing worse than expected, it could be caused by the referring link text or placement rather than the quality of the referrer. Conversely, an unexpected spike in referrals from a certain page could be either good or bad depending on the content of the referring page.

In the same way that the referrer metric helps us to assess referrer effectiveness, keyword analysis helps us to measure the referrer value of referring search engines and shows which keywords have brought in the most traffic. By analyzing the keywords visitors use to find a page, we can determine what visitors expect to gain from the Website and use that information to better tailor the Website to their needs. It is also important to consider the quality of keywords. Keyword quality is directly proportional to revenue and can be determined by comparing keywords with visitor path and visit. Good keywords will bring quality traffic and more income to the site.

5.8   ERRORS

Errors are the final metric. Tracking errors has the obvious benefit of being able to identify and fix any errors in the Website, but it is also useful to observe how visitors react to these errors. The fewer visitors who are confused by errors on a Website, the less likely visitors are to exit the site because of an error.

5.9   CONCLUSION

Once we understand these eight fundamental metrics, we can begin to develop a coherent Web analytics strategy.

*   *   *   *

CHAPTER 6

Web Analytics Strategy

Through unobtrusive transaction logs and page tags we can gather a massive amount of data about user–system interaction, and by employing fundamental metrics we can evaluate human behavior within the interactional context. In order to gain the most from these massive datasets, however, we must strategically select and employ the fundamental metrics in relation to KPIs. For example, by collecting various Web analytics metrics, such as number of visits and visitors and visit duration, we can develop KPIs, thereby creating a versatile analytic model that measures several metrics against each other to define visitor trends [19, 111]. One primary concern in developing a coherent Web analytic strategy is understanding the relationships among the foundational metrics and KPIs.

6.1 KEY PERFORMANCE INDICATORS

KPIs provide an in-depth picture of visitor behavior on a site. This information allows organizations to align their Websites' goals with their business goals for the purpose of identifying areas of improvement, promoting popular parts of the site, testing new site functionality, and ultimately increasing revenue. This section covers the most common metrics, different ways for gathering metrics, methods for utilizing KPIs, best key practices, and the selection criteria for choosing the right Web analytics tool. In brief, this section describes the overall process of and provides advice for Web analytics integration, and discusses the future of Web analytics.

Before beginning this discussion, we need to clarify the exact meaning of frequently used terms. For our purpose, we will use the following definitions [20]:

- **Measurement:** In the most general terms, measurement can be regarded as the assignment of numbers to objects (or events or situations) in accord with some rule (measurement function). The property of the objects that determines the assignment according to that rule is called magnitude, the measurable attribute; the number assigned to a particular object is called its measure, the amount or degree of its magnitude. Importantly, the rule defines both the magnitude and the measure.

- **Web Page Significance:** Significance metrics formalize the notions of "quality" and "relevance" of Web pages with respect to users' information needs. Significance metrics are employed to rate candidate pages in response to a search query and to influence the quality of search and retrieval on the Web.

- **Usage Characterization:** Patterns and regularities in the way users browse Web resources can provide invaluable clues for improving the content, organization, and presentation of Websites.

Usage characterization metrics measure user behavior for this purpose.

6.2 WEB ANALYTICS PROCESS

Although we can collect metrics from a Website, we must be mindful of how we gather them and how we use them to select and filter information. Effective design and use of Web analytics requires us to do so [24]. How can the strategic use of Web analytics help improve an organization? To answer this, the WAA provides nine key best practices to follow when analyzing a Website [101]. Figure 6.1 outlines this process.

6.2.1 Identify Key Stakeholders

The first step in the process of Web analytics is to identify the key stakeholders, meaning anyone who holds an interest in the Website. This designation includes management, site developers, visitors, and anyone else who creates, maintains, uses, or is affected by the site. In order for the Website

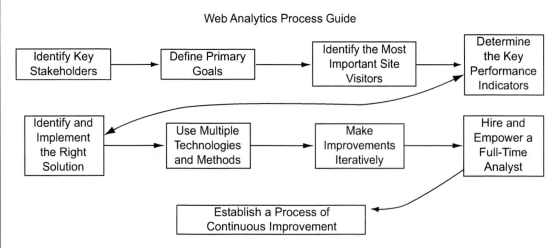

FIGURE 6.1: The process of Web analytics [101].

to be truly beneficial, it must integrate input from all major stakeholders. Involving people from different parts of the company also makes it more likely that they will embrace the Website as a valuable tool.

6.2.2 Define Primary Goals for Your Website

Knowing the key stakeholders and understanding their primary goals assists us in determining the primary goals of the Website. Such goals could include increasing revenue, cutting expenses, and increasing customer loyalty [101]. After defining the Website's goals, it is important to prioritize them in terms of how the Website can most benefit the company. While seemingly simplistic, the task can be challenging. Political conflict between stakeholders and their individual goals as well as inaccurate assumptions they may have made while determining their goals can derail this process. Organizational leaders may need to be consulted to manage the conflict, but we can assist by keeping the discussion focused on the overarching organizational goals and Website capabilities. By going through this process, a company can minimize conflict among competing goals and maintain relationships with various stakeholders.

6.2.3 Identify the Most Important Site Visitors

One group of stakeholders that is critical to Websites is visitors. According to Sterne, corporate executives categorize their visitors in terms of importance. Most companies classify their most important visitors as ones who either visit the site regularly, stay the longest on the site, view the most pages, purchase the most goods or services, purchase goods most frequently, or spend the most money [143]. There are three types of customers: (1) customers a company wants to keep who have a high current value and high future potential, (2) customers a company wants to grow who can either have a high current value and low future potential or low current value and high future potential, and (3) customers a company wants to eliminate who have a low current value and low future potential. The most important visitor to a Website, however, is the one who ultimately brings in the most revenue. Categorizing visitors as customer types enables us to consider their goals more critically. What improvements can we make to the Website in order to improve visitors'/customers' browsing experiences and intentionally to grow more visitors and revenue?

6.2.4 Determine the Key Performance Indicators

To assist organizations in determining how to improve their Website, Web analytics offers the strategic use and monitoring of KPIs. This involves picking the metrics that will be most beneficial in improving the site and eliminating the ones that will provide little or no insight into its goals. The

Website type—commerce, lead generation, media/content, or support/self-service—plays a key role in which KPIs are most effective for analyzing site traffic.

6.2.5   Identify and Implement the Right Solution

After the KPIs have been defined, the next step is identifying the best Web analytics technology to meet the organization's specific needs. The most important things to consider are the budget, software flexibility, ease of use, and compatibility between the technology, and the metrics. McFadden wisely suggests that a pilot test of the top two vendor choices can ease the decision making [101]. We will expand on this topic in the next section.

6.2.6   Use Multiple Technologies and Methods

Web analytics is not the only method available for improving a Website. To achieve a more holistic view of a site's visitors, we can also use tools such as focus groups, online surveys, usability studies, and customer services contact analysis [101].

6.2.7   Make Improvements Iteratively

While we all may want to address identified Website problems quickly, we should make only gradual improvements. This incremental method allows us to monitor whether a singular change is an improvement and how much of an improvement. Importantly, it also provides us with the opportunity to assess the change within the overall Website context and observe what systemic issues the small change may have created.

6.2.8   Hire and Empower a Full-Time Analyst

Monitoring and assessing a Website's effectiveness is a complex and compounding process. For these reasons, it is important to have one person who consistently analyzes the data generated and determine when, how, and why new information is needed. According to the WAA, a good analyst understands organizational needs (which means communicating well with stakeholders); has knowledge of technology and marketing; has respect, credibility, and authority; and is already a company employee. Although it may seem like hiring a full-time analyst is expensive, many experts agree that the return on revenue should be more than enough compensation [101].

6.2.9   Establish a Process of Continuous Improvement

Once the Web analysis process is established, continuous evaluation is paramount. This means reviewing the goals and metrics and monitoring new changes and features as they are added to the Website. It is important that the improvements are adding value to the site and meeting expectations.

6.3   CHOOSING A WEB ANALYTICS TOOL

Once one decides what is wanted from Web analysis, it is time to find the right tool. Kaushik outlines ten important questions to ask Web analytics vendors [79]:

- **What is the difference between your tool and free Web analytics tools?** Since the company who owns the Website will be paying money for a service, it is important to know why that service is better than free services (e.g., Google Analytics). Look for an answer that outlines the features and functionality of the vendor. Do not look for answers about increased costs because of privacy threats or poor support offered by free analytics tools.

- **Do you offer a software version of your tool?** Generally, a business will want to look for a tool that is software based and that can run on their own servers. If a tool does not have a software version but plans to make one in the future, it shows insight into how prepared they are to offer future products if there is interest.

- **What methods do you use to capture data?** As stated earlier, there are two main ways to capture visitor data from a Website: log files and page tagging. Ideally, we prefer a vendor that offers both, but what they have used in the past is also important. Because technology is constantly changing, we want a company that has a history of keeping up with and perhaps even anticipating market changes and that has addressed these dynamics through creative solutions.

- **Can you help me calculate the total cost of ownership for your tool?** The total cost of ownership for a Web analytics tool depends on the specific company, the systems they have in place, and the pricing of the prospective Web analytics tool. In order to make this calculation, we must consider the following:
  1. Cost per page view
  2. Incremental costs (i.e., charges for overuse or advanced features)
  3. Annual support costs after the first year
  4. Cost of professional services (i.e., installation, troubleshooting, or customization)
  5. Cost of additional hardware we may need
  6. Administration costs (which includes the cost of an analyst and any additional employees we may need to hire)

- **What kind of support do you offer?** Many vendors advertise free support, but it is important to be aware of any limits that could incur additional costs. It is also important to note how extensive their support is and how willing they are to help.

- **What features do you provide that will allow me to segment my data?** Segmentation allows companies to manipulate their data. Look for the vendor's ability to segment the data after it is recorded. Many vendors use JavaScript tags on each page to segment the data as it is captured, meaning that the company has to know exactly what it wants from the data before having the data itself; this approach is less flexible.

- **What options do I have to export data into our system?** It is important to know who ultimately owns and stores the data and whether it is possible to obtain both raw and processed data. Most vendors will not provide companies with the data exactly as they need it, but it is a good idea to realize what kind of data is available before making a final decision.

- **Which features do you provide for integrating data from other sources into your tool?** Best practice, as noted previously, recommends using multiple technologies and methods in order to inform decision making. If a company has other data it wants to bring to the tool (such as survey data or data from an ad agency), then it is important to know whether this information can be integrated into the vendor's analytic tool.

- **What new features are you developing that would keep you ahead of your competition?** Not only will the answer to this question tell how much the vendor has thought about future functionality, but it will also show how much they know about their competitors. If they are trying to anticipate changes and market demands, then they should be well informed about their competition.

- **Why did you lose your last two clients? Who are they using now?** The benefits of this question are obvious—by knowing how they lost previous business, the business can be confident that it has made the right choice.

6.4   CONCLUSION

With an effective Web analytics strategy in place, we can turn our attention to understanding user behaviors and identifying necessary or potentially beneficial system improvements. In practice, this is rarely the end. Web analytics strategy typically supports some overarching goals.

· · · ·

CHAPTER 7

Web Analytics as Competitive Intelligence

In order to get the most out of Web analytics, we must first effectively choose which metrics to analyze and then combine them in meaningful ways [19].This means knowing the Website's business goals and then determining which KPIs will provide the most insight for these business goals.

7.1 DETERMINING APPROPRIATE KEY PERFORMANCE INDICATORS

To determine appropriate KPIs, one must know your business goals. Every company has specific business goals. Every part of the company works together to achieve them, and the company Website is no exception. In order for a Website to be beneficial, information gathered from its visitors must not merely show what has happened in the past, but it must also be able to improve the site for future visitors. The company must have clearly defined goals and must use this information to support strategies that will help it achieve those goals.

For a Website, the first step is making sure the data collected from the site is actionable. According to the WAA [101], in order for a company to collect actionable data, it must meet three criteria [144]:

- the business goals must be clear,
- technology, analytics, and the business must be aligned, and
- the feedback loop must be complete.

There are many possible methods for meeting these criteria. One is Alignment-Centric Performance Management [14]. This approach goes beyond merely reviewing past customer trends to carefully selecting a few key KPIs based on future business objectives. Even though a wealth of metrics is available from a Website, not all of the metrics are relevant to a company's needs. Moreover, reporting large quantities of data is overwhelming, so it is important to look at metrics in context and use them to create KPIs that focus on outcome rather than activity. For example, a customer service Website might view the number of emails responded to on the same day they were sent as a measurement of customer satisfaction. A better way to measure customer satisfaction, however, might be to survey the customers on their experience. Although this measurement is subjective, it is a better representation of customer satisfaction because even if a customer receives a response the same day he or she sent an email, the customer may still be dissatisfied with the service experience [14].

Following "The Four M's of Operational Management," as outlined by Becher [14], can facilitate effective selection of KPIs:

- **Motivate**: ensure that goals are relevant to everyone involved.
- **Manage**: encourage collaboration and involvement for achieving these goals.
- **Monitor**: once selected, track the KPIs and quickly deal with any problems that may arise.
- **Measure**: identify the root causes of problems and test any assumptions associated with the strategy.

By carefully choosing a few, quality KPIs to monitor and making sure everyone is involved with the strategy, we can more easily align a Website's goals with the company's goals because the information is targeted and stakeholders are actively participating.

Another method for ensuring actionable data is Online Business Performance Management (OBPM) [132]. This approach integrates business tools with Web analytics to help companies make better decisions quickly in an ever-changing online environment where customer data is stored in a variety of different departments. The first step in this strategy is gathering all customer data in a central location and condensing it so that the result is all actionable data. Once this information is in place, the next step is choosing relevant KPIs that align with the company's business strategy and then analyzing expected versus actual results [132].

In order to choose the best KPIs and measure the Website's performance against the goals of a business, there must be effective communication between senior executives and online managers. The two groups should work together to define the relevant performance metrics, the overall goals for the Website, and the performance measurements. This method is similar to Alignment-Centric Performance Management in that it aims to aid integration of the Website with the company's business objectives by involving major stakeholders. The ultimate goals of OBPM are increased confidence, organizational accountability, and efficiency [132].

Of course, one must identify KPIs based on the Website type. Unlike metrics, which are numerical representations of data collected from a Website, KPIs are tied to a business strategy and are usually measured by a ratio of two metrics. By choosing KPIs based on the Website type, a busi-

WEBSITE TYPE	KPIS
Commerce	• Average order value • Average visit value • Bounce rate • Conversion rates • Customer loyalty
Content/Media	• New visitor ratio • Page depth • Returning visitor ratio • Visit depth
Lead Generation	• Bounce rate • Conversion rates • Cost per lead • Traffic concentration
Support/Self-service	• Bounce rate • Customer satisfaction • Page depth • Top internal search phrases

TABLE 7.1: The four types of Websites and examples of associated KPIs [101].

ness can save both time and money. Although Websites can have more than one function, each site belongs to at least one of the four main categories: commerce, lead generation, content/media, and support/self-service [101]. Table 7.1 shows common KPIs for each Website type.

We discuss each Website type and related KPIs below.

7.1.1 Commerce

The goal of a commerce Website is to get visitors to purchase goods or services directly from the site, with success gauged by the amount of revenue the site brings in. According to Peterson, "commerce analysis tools should provide the 'who, what, when, where, and how' for your online purchasers" [118, p. 92]. In essence, the important information for a commerce Website is to answer the following questions: Who made (or failed to make) a purchase? What was purchased? When were purchases made? From where are customers coming? How are customers making their purchases? The most valuable KPIs used to answer these questions are conversion rates, average order value, average visit value, customer loyalty, and bounce rate [101]. Other metrics to consider with a commerce site are which products, categories, and brands are sold on the site and an internal site product search that could signal navigation confusion or a new product niche [118].

A conversion rate is the number of users who perform a specified action divided by the total of a certain type of visitor (i.e., repeat visitors, unique visitors, etc.) over a given period. Types of conversion rates will vary by the needs of the businesses using them, but two common conversion rates for commerce Websites are the order conversion rate (the percent of total visitors who place an order on a Website) and the checkout conversion rate (the percent of total visitors who begin the checkout process). There are also many methods for choosing the group of visitors on which to base the conversion rate. For example, businesses may want to filter visitors by excluding visits from robots and Web crawlers [5], or they may want to exclude the traffic that "bounces" from the Website or (a slightly trickier measurement) the traffic that is determined not to have intent to purchase anything from the Website.

Commerce Websites commonly have conversion rates of around 0.5%, but generally good conversion rates will fall in the 2% range, depending on how a business structures its conversion rate [41]. Again, the ultimate goal is to increase total revenue. According to eVision, a search engine marketing company, for each dollar a company spends on improving this KPI, there is 10 to 100 multiple return [39]. The methods a business uses to improve the conversion rate (or rates), however, are different depending on which target action that business chooses to measure.

Average order value is a ratio of total order revenue to number of orders over a given period. This number is important because it allows the analyst to derive a cost for each transaction. There are several ways for a business to use this KPI to its advantage. One way is to break down the average order value by advertising campaigns (i.e., email, keyword, banner ad, etc.). In this way, a business can see which campaigns are bringing in the best customers and then opt to spend more effort refining strategies in those areas [119]. Overall, however, if the cost of making a transaction is greater than the amount of money customers spend for each transaction, then the site is not fulfilling its goal. There are two main ways to correct this. The first is to increase the number of products customers order per transaction, and the second is to increase the overall cost of purchased products. A good technique for achieving either of these goals is product promotions [101], but many factors influence how and why customers purchase what they do on a Website. These factors are diverse and can range from displaying a certain security image on the site [97] to updating the site's internal search [161]. Like many KPIs, improvement ultimately comes from ongoing research and a small amount of trial and error.

Another KPI, average visit value, measures the total number of visits to the total revenue and essentially informs businesses about the traffic quality. It is problematic for a commerce site when, even though it may have many visitors, each visit generates only a small amount of revenue. In that case, increasing the total number of visits would likely increase profits only marginally. The average visit value KPI is also useful for evaluating the effectiveness of promotional campaigns. If the average visit value decreases after a specific campaign, it is likely that the advertisement is not attracting quality traffic to the site. Another less common factor in this situation could be broken links or a confusing layout in a site's "shopping cart" area. A business can improve the average visit value by using targeted advertising and employing a layout that reduces customer confusion.

One way to assess customer quality is to identify customer loyalty. This KPI is the ratio of new to existing customers. Many Web analytics tools measure this using visit frequency and transactions, but there are several important factors in this measurement including the time between visits [100]. Customer loyalty can even be measured simply with customer satisfaction surveys [133]. Loyal customers will not only increase revenue through purchases but also through referrals, potentially limiting advertising costs [123].

Yet another KPI that relates to customer quality is bounce rate. Essentially, bounce rate measures how many people arrive at a homepage and leave immediately. Two scenarios generally qualify as a bounce. In the first scenario, a visitor views only one page on the Website. In the second scenario, a visitor navigates to a Website but only stays on the site for 5 seconds or less [6]. This could be due to several factors, but in general visitors who bounce from a Website are not interested in the content. Like average order value, this KPI helps show how much quality traffic a Website receives. A high bounce rate may reflect counterintuitive site design or misdirected advertising.

7.1.2   Lead Generation

The goal for a lead generation Website is to obtain user contact information in order to inform them of a company's new products and developments and to gather data for market research; these sites primarily focus on products or services that cannot be purchased directly online. Examples of lead generation include requesting more information by mail or email, applying online, signing up for a newsletter, registering to download product information, and gathering referrals for a partner site [23]. The most important KPIs for lead generation sites are conversion rates, cost per lead (CPL), bounce rate, and traffic concentration [101].

Similar to commerce Website KPIs, a conversion rate is the ratio of total visitors to the amount of visitors who perform a specific action. In the case of lead generation Websites, the most common conversion rate is the ratio of total visitors to leads generated. The same visitor filtering techniques mentioned in the previous section can be applied to this measurement (i.e., filtering out robots and Web crawlers and excluding traffic that bounces from the site). This KPI is an essential tool in analyzing marketing strategies. Average lead generation sites have conversion rates ranging from 5–6% to 17–19% for exceptionally good sites [46]. Conversion rates that increase after the implementation of a new marketing strategy indicate that the campaign was successful. Decreases in conversion rates indicate that the campaign was not effective and probably needs to be reworked.

Another way to measure marketing success is CPL, which is the ratio of total expenses to total number of leads or how much it costs a company to generate a lead; a more targeted measurement of this KPI would be the ratio of total marketing expenses to total number of leads. A good way to measure the success of this KPI is to make sure that the CPL for a specific marketing campaign is less than the overall CPL [155]. Ideally, the CPL should be low, and well-targeted advertising is usually the best way to achieve this.

Lead generation bounce rate is the same measurement as the bounce rate for commerce sites. This KPI measures visitor retention based on total number of bounces to total number of visitors (a bounce is characterized by a visitor entering the site and immediately leaving). Lead generation sites differ from commerce sites in that they may not require the same level of user interaction. For example, a lead generation site could have a single page where users enter their contact information. Even though they only view one page, the visit is still successful if the Website is able to collect the user's information. In these situations, it is best to base the bounce rate solely on time spent on the site. As with commerce sites, the best way to decrease a site's bounce rate is to increase advertising effectiveness and decrease visitor confusion.

The final KPI is traffic concentration, or the ratio of the number of visitors to a certain area in a Website to total visitors. This KPI shows which areas of a site have the most visitor interest. For lead generation Websites, it is ideal to have a high traffic concentration on the page or pages where users enter their contact information.

7.1.3   Content/Media

Content/media Websites focus mainly on advertising, and the main goal of these sites is to increase revenue by keeping visitors on the Website longer and to keep visitors coming back to the site. In order for these types of sites to succeed, site content must be engaging and frequently updated. If content is only part of a company's Website, the content used in conjunction with other types of pages can be used to draw in visitors and provide a way to immerse them in the site. The main KPIs are visit depth, returning visitors, new visitor percentage, and page depth [101].

Visit depth (also referred to as depth of visit or path length) is the measurement of the ratio between page views and unique visitors, or how many pages a visitor accesses each visit. As a general rule, visitors with a higher visit depth interact more with the Website. If visitors are only viewing a few pages per visit, then they are not engaged, indicating that the site's effectiveness is low. One way to increase a low average visit depth is by creating more targeted content that would be more interesting to the Website's target audience. Another strategy could be increasing the site's interactivity to encourage the users to become more involved with the site and to motivate them to return frequently.

Unlike the metric of simply counting the number of returning visitors on a site, the returning visitor KPI is the ratio of unique visitors to total visits. A factor in customer loyalty, this KPI measures the effectiveness of a Website attracting repeat visitors. A lower ratio for this KPI is best because it indicates more repeat visitors and more visitors who are interested in and trust the content of the Website. If this KPI is too low, however, it might signal problems in other areas such as a high bounce rate or even click fraud. Click fraud occurs when a person or script is used to generate visits to a Website without having genuine interest in the site. According to a study by Blizzard Internet Marketing, the average for returning visitors to a Website is 23.7% [157]. As with many of the other KPIs for content/media Websites, the best way to improve the returning visitor rate is by having quality content and encouraging interaction with the Website.

Content/media sites are also interested in attracting new visitors, and the new visitor ratio compares new visitors with unique visitors to determine if a site is attracting new people. New visitors can be brought to the Website in a variety of different ways, so a good way to increase this KPI is to try different marketing strategies to determine which campaigns bring the most (and the best) traffic to the site. When using this KPI, we must keep the Website's goal in mind. Specifically, is the Website intended more to retain or to attract customers? When measuring this KPI, the age of the Website plays a role—newer sites will want to attract new people. As a rule, however, the new visitor ratio should decrease over time as the returning visitor ratio increases. The final KPI for content/media sites is page depth. This is the ratio of page views for a specific page and the number of unique visitors to that page. This KPI is similar to visit depth, but its measurements focus more on page popularity. Average page depth can indicate interest in specific areas of a Website over time and measure whether the interests of the visitors match the goals of the Website. If one particular page on a Website has a high page depth, then that page is of particular interest to visitors. An example of a page in a Website expected to have a higher page depth would be a news page. Information on a news page is updated constantly so that, while the page is still always in the same location, the content of that page is constantly changing. If a Website has high page depth in a relatively unimportant part of the site, it may signal visitor confusion with navigation in the site or an incorrectly targeted advertising campaign.

7.1.4   Support/Self-Service

Websites offering support or self-service are interested in helping users find specialized answers for specific problems. The goals for this type of Website are increasing customer satisfaction and decreasing call center costs; it is more cost-effective for a company to have visitors find information through its Website than it is to operate a call center. The KPIs of interest are visit length, content depth, and bounce rate. In addition, other areas to examine are customer satisfaction metrics and top internal search phrases [101].

Page depth for support/self-service sites is the same measurement as page depth content/media sites, namely the ratio of page views to unique visitors. With support/self-service sites, however, high page depth is not always a good sign. For example, a visitor viewing the same page multiple times may show that the visitor is having trouble finding helpful information on the Website or even that the information the visitor is looking for does not exist on the site. The goal of these types of sites is to help customers find what they need as quickly as possible and with the least amount of navigation through the site. The best way to keep page depth low is to keep visitor confusion low.

As with the bounce rate of other Website types, the bounce rate for support/self-service sites reflects ease of use, advertising effectiveness, and visitor interest. A low bounce rate means that quality traffic is coming to the Website and deciding that the site's information is potentially useful. Poor advertisement campaigns and poor Website layout will increase a site's bounce rate.

Customer satisfaction deals with how users rate their experience on a site and is usually collected directly from the visitors (not from log files), either through online surveys or through satisfaction ratings. Although it is not a KPI in the traditional sense, gathering data directly from visitors to a Website is a valuable tool for figuring out exactly what visitors want. Customer satisfaction measurements can deal with customer ratings, concern reports, corrective actions, response time, and product delivery. Using these numbers, we can compare the online experience of the Website's customers with the industry's average and make improvements according to visitors' expressed needs.

Site navigation is important to visitors, and top internal search phrases, which apply only to sites with internal search capabilities, can be used to measure what information customers are most interested in that can inform site navigation improvements. Moreover, internal search phrases can be used to direct support resources to the areas generating the most user interest, as well as to identify which parts of the Website users may have trouble accessing. Other problems may also become obvious. For example, if many visitors are searching for a product not supported on the Website, then this may indicate that the site's marketing campaign is ineffective.

Regardless of Website type, the KPIs listed above are not the only KPIs that can prove useful in analyzing a site's traffic, but they provide a good starting point. The main thing to remember is that no matter what KPIs a company chooses to use, they must be aligned with its goals, and more KPIs do not necessarily mean better analysis: quality is more important than quantity.

7.2   CONCLUSION

Any organization, business, or Website must start with clearly defined goals because they are essential for any successful strategy. With a clearly defined and understood strategy, we can then plan and implement the tactics necessary for executing this strategy. These tactics are based on KPIs—which are the measures and metrics of performance. As such, KPIs are the foundation for any Web goal achievement.

CHAPTER 8

Supplementary Methods for Augmenting Web Analytics

While the relatively unobtrusive methods of data collection that we have discussed thus far are very valuable, proponents of using transaction logs for Web analysis typically admit that the method has shortcomings [66, 91], as do all methodological approaches. These shortcomings include failing to understand the affective, situational, and cognitive aspects of system users. Therefore, we must look to other methods in order to address some of these shortcomings and limitations [124]. Fortunately, the Web and other information technologies provide a convenient means for employing surveys and survey research for such a purpose.

As an overview, we discuss surveys and laboratory studies as viable alternative methods for Web log analysis, and then present a brief review of survey and laboratory research literature, with a focus on the use of surveys and laboratory studies for Web-related research. The section then identifies the steps in implementing survey research and designing a survey instrument and a laboratory study.

8.1 SURVEYS

Survey research is a method for gathering information by directly asking respondents about some aspect of themselves, others, objects, or their environment. As a data collection method, survey instruments are very useful for a variety of research designs. For example, researchers can use surveys to describe current characteristics of a sample population and to discover the relationship among variables. Surveys gather data on respondents' recollections or opinions; therefore, surveys provide an excellent companion method for Web analytics that typically focus exclusively on actual behaviors of participants [125].

After reviewing some studies that have used surveys for Web research, we will discuss how to select, design, and implement survey research.

8.1.1 Review of Appropriate Survey Literature

Although surveys have been used for hundreds of years, the Web provides a remarkable channel for the use of surveys to conduct data collection [75]. Many of these Internet surveys have focused on demographical aspects of Web use over time [83] or one particular Website feature [150]. Treiblma-ier [148] presents an extensive review of the use of surveys for Website analysis.

Survey respondents may include general Web users or samples from specific populations. For example, Huang [61] surveyed users of continuing education programs. Similarly, Jeong et al. [76] surveyed travel and hotel shoppers, and Kim and Stoel [86] surveyed female shoppers who have purchased apparel online.

For academic researchers, a convenience sample of students is often used to facilitate survey studies, including the users of Web search engines [139]. McKinney et al. [103] used both under-graduate and graduate students as their sample examining Website use. The major advantages of using students that are often cited include a homogeneous sample, access [62], familiarity with the Internet [67], and creation of experimental settings [130]. There are concerns in generalizing these results [1], most notably for Websites and services where students have limited domain or system knowledge [86, 89]. However, as a sample of demographic slice of the Web population, students appear to be a workable convenience sample with results from studies with students [cf. Refs. 67, 84] similar to those using more rigorous sampling methods [cf. Refs. 51, 83]. Organizations such as the Pew Research Center's Internet and American Life Project use random samples of the U.S. Web population for their surveys [125].

For the Web, the most common type of survey instruments are electronic or Web surveys. Jansen et al. [75] define an electronic survey as "one in which a computer plays a major role in both the delivery of a survey to potential respondents and the collection of survey data from actual re-spondents" (p. 1). Several researchers have examined electronic survey approaches, techniques, and instruments with respect to methodological issues associated with their use [33, 35, 40, 43, 90, 145]. There have been mixed research results concerning the benefits of electronic surveys [85, 104, 142, 149]. However, researchers generally agree that electronic surveys offer faster response times and decreased costs. The electronic and Web-based surveys allow for a nearly instantaneous data collec-tion into a backend database, which reduces potential errors caused by manual transcription.

Regardless of which delivery method is used, survey research requires a detailed project plan-ning approach.

8.1.2 Planning and Conducting a Survey

Although conducting a survey may appear to be an easy task, the reality is quite the opposite. Suc-cessful survey research requires detailed planning. The goal of any survey is to shed insight into how the respondents perceive themselves, their environment, their context, their situation, their behaviors, or their perceptions of others.

To execute a survey, the researcher must identify the content area, construct the survey instrument, define the population, select a representative sample, administer the survey instrument, analyze and interpret the results, and then communicate the results. While these steps are somewhat linear, they also overlap and may require several iterations. A 10-step survey research process is illustrated in Table 8.1, based on a process outlined in Graziano and Raulin [45].

Steps 1 and 2: Determine the specific information desired and define the population to be studied. The information being sought and the population to be studied are the first tasks of the survey researcher. The goals of the survey research will determine both the information being sought and the target population. Additionally, the goals will drive both the construction and administration of the survey. If we use a survey to supplement ongoing Web log analysis, then these decisions will follow the established parameters.

Step 3: Decide how to administer the survey. There are many possibilities for administering a survey, ranging from face-to-face (i.e., an interview), to pen and paper, to the telephone (i.e., phone survey), to the Web (i.e., electronic survey). A survey can also be a mixed mode survey, combining

TABLE 8.1: Process for conducting and implementing a survey [125].

STEP	ACTIONS
1	Determine the specific information desired
2	Define the population to be studied
3	Decide how to administer the survey
4	Design a survey instrument
5	Pretest the survey instrument with a subsample
6	Select a sampling approach and representative sample
7	Administer the survey instrument to the sample
8	Analyze the data
9	Interpret the findings
10	Communicate the results to the appropriate audience

more than one of these approaches. The exact method selected really depends on the answers to steps one and two (i.e., what information is needed and what population is studied). Used in conjunction with Web analytics, surveys can be conducted either before or after a laboratory study. A survey can also be used to gain insight into the demographics of the wider Web population.

Step 4: Designing a survey instrument. Developing a survey instrument involves several steps. The researcher must determine what questions to ask, in what form, and in what order. The researcher must construct the survey so that it adequately gathers the information being sought. A basic rule of survey research is that the instrument should have a clear focus and should be guided by the research questions or hypotheses of the overall study. This implies that survey research is not well suited to early exploratory research because it requires some orderly expectations and focus from the researcher.

Step 5: Pretest the survey instrument with a subsample. Once the researcher has the survey instrument ready and refined, the researcher must pilot test the survey instrument. In this respect, a survey instrument is like developing a system artifact, where a system is beta-tested before wider deployment. Generally, one conducts the pilot test on a sample that represents the population being studied, after which the researcher may (generally, will) refine the survey instrument further. Depending on the extent of the changes, the survey instrument may require another pilot test.

Step 6: Select a sampling approach and representative sample. Selecting an adequate and representative sample is a critical and challenging factor in survey research. The population for survey study is the larger group about or from whom the researcher desires to obtain information. From this population, we need to survey a representative sample. If we are administering a survey to the respondents of a laboratory study, the representativeness is not a problem because the respondents are the representative sample. Selecting a representative sample of Web users, however, requires careful planning.

Whenever we use a sample as a basis for generalizing to a population, we engage in an inductive inference from the specific sample to the general population. In order to have confidence in inductive inferences from sample to population, the researcher must carefully choose the sample to represent the overall population. This is especially true for descriptive research, where the researcher wishes to describe some aspect of a population that may depend on demographic characteristics. In other cases, such as verifying the application of universal theoretical constructs, for example, Zipf's Law [164], sampling is not as important since these universal constructs should apply to everyone within the population.

Sampling procedures typically fall into three classifications:

- **Convenience sampling** (i.e., selecting a sample with little concern for its representativeness to some overall population),

- **Probability sampling** (i.e., selecting a sample where each respondent has some known probability of being included in the sample), and
- **Stratified sampling** (i.e., selecting a sample that includes representative samples of each subgroup within a population).

Step 7: Administer the survey instrument to the sample. For actually gathering the survey data, the researcher must determine the most appropriate manner to administer the survey instrument. Many surveys are administered via the Web or electronically, as the Web offers substantial benefits in its easy access to a wide population sample. Additionally, administering a survey electronically, even in a laboratory study, has significant advantages in terms of data preparation for analysis. The survey can be administered once to a cross-sectional portion of the population, or it can be administered repeatedly to the same sample population.

Step 8: Analyze the data. Once the data is gathered, we must determine the appropriate method for analysis. The appropriate form of analysis is dependent on the research questions, hypotheses, or types of questions used in the survey instruments. The available approaches are qualitative, quantitative, and mixed methods.

Step 9: Interpret the finding. Like many research results, the interpretation of survey data can be somewhat subjective. When results are in question, it may point to the need for further research. One of the best aids in interpreting results is the literature review. What have results from previous work pointed out? Are these results in line with those previous researches? Or do the results highlight something new?

Step 10: Communicate the results to the appropriate audience. Finally, the results of any survey research must be packaged for the intended audience. For academic purposes, this may mean a scholarly paper or presentation. For commercial organizations, this may mean a white paper for system developers or marketing professionals.

Each of these steps can be challenging. However, designing a survey instrument (e.g., Steps 4 and 5) can be the most difficult aspect of the survey research. We address this development in more detail in the following section.

8.1.3   Design a Survey Instrument

Before designing a survey instrument, the researcher must have a clear understanding of the type of data desired and must keep the instrument focused on that area. The key to obtaining good data via a survey is to develop a good survey instrument that is based on the research questions. The researcher should develop a set of objectives with a clear list of all needed data. The research goals and list of needed data will serve as the basis for the questions on the survey instrument.

What is your gender?	Which features of Instant Messaging programs do you find most useful when it comes to sharing information with teammates?
a. Male	a. Real-Time Chat
b. Female	b. File Sharing
	c. Chat logs
	d. None

FIGURE 8.1: Examples of multiple-choice questions.

A survey instrument is a data collection method that presents a set of questions to a respondent. The respondent's responses to the questions provide the data sought by the researcher. Although seemingly simple, it can be very difficult to develop a set of questions for a survey instrument. Some general guidelines for developing survey instruments are [113]:

- **State on the survey instrument the research goal:** At the top of the survey instrument, include a very brief statement explaining the purpose of the survey and assuring respondents of their anonymity.

- **Provide instructions for completing the survey instrument:** To assist in ensuring that survey results are valid, include instructions on how to respond to questions on the survey instrument. Generally, there is a short introductory set of instructions at the top of the survey instrument. Provide additional instructions for specific questions if needed.

- **Place questions concerning personal information at the end of the survey:** Demographic information is often necessary for survey research. Place these questions at the end of the survey. Providing personal data may annoy some respondents, resulting in incomplete or inaccurate responses to the survey instrument.

- **Group questions on the instrument by subject:** If the survey instrument has more than 10 or so questions, the questions need to be grouped by some classification method. Generally, grouping the questions by subject is a good organization method. If the instrument has

On a scale of 1–7, would you search individually or together with your workmates if you do not know anything about the problem?

	Individual						Collaborate
	1	2	3	4	5	6	7
	*	*	*	*	*	*	*

FIGURE 8.2: Example of a rating question.

On a scale of 1–5 (1—never used, 5 — use every day), rank the following items on how expe-rienced are you with using the following communication/collaboration applications for group projects?

    a. _____ Email
    b. _____ Instant messaging
    c. _____ Face-to-face meetings
    d. _____ Telephone
    e. _____ Others (please elaborate)

FIGURE 8.3: Example of a ranking question.

multiple groups of questions, each group should have a heading identifying the grouping. Grouping questions allows the respondents to focus their responses around the central theme of the group of questions.

- **Present each question and type of question in a consistent structure:** A consistent struc-ture makes it much simpler for respondents and increases the likelihood of valid data. Explain the proper method for responding to each question and ensure that the response methods for similar questions are consistent throughout the instrument.

There are three general categories of survey questions, namely multiple-choice, Likert-scale, and open-ended questions.

Multiple-choice questions. Multiple-choice questions have a closed set of response items for the respondents to select. Multiple-choice questions are useful when we have a thorough under-standing of the range of possible responses (see Figure 8.1).

The items for multiple-choice questions must cover all possible alternatives that the re-spondents might select, and each of the items must be unique (i.e., they must not overlap). Since

As part of your project, I believe that you must have confronted a situation when you did not really know how to proceed in order to solve a problem or perform a task on the Web.

(a) Can you speak about a specific instance of your project work in which you were uncer-tain as to how to proceed?

_____

_____

FIGURE 8.4: Example of an open-ended question.

Which features of Instant Messaging programs do you find most useful when it comes to sharing information with teammates?

    a. Real-Time Chat
    b. File Sharing
    c. Chat logs
    d. Others _____
    e. None

FIGURE 8.5: Example of a partially structured question.

presenting all possible alternatives is a difficult task, we normally include a general catch-all item
(e.g., *None of the above* or *Don't know*) at the end of a list of item choices. This approach helps
improve the accuracy of the data collected.

    Likert-scale questions. With Likert-scale questions, the items are arranged as a continuum
with the extremes generally at the endpoints. Likert-scale questions may have respondents indicate
the degree to which they agree with a statement (see Figure 8.2) or rank a list of items (see Figure
8.3).

    Open-ended questions. As Figure 8.4 demonstrates, open-ended questions have no list of
items for the respondent to choose from.

    Open-ended questions are best for exploring new ideas or when the researcher does not know
any of the expected responses. As such, the open-ended questions are great for qualitative research.
The disadvantages to using open-ended questions are that it can be much more time consuming and
difficult to analyze the data because each question must be coded into order to derive variables.

    If we have a partial list of possible responses, one can create a partially open-ended question
(see Figure 8.5).

8.2   LABORATORY STUDIES

A laboratory study is research conducted in a laboratory setting to investigate aspects that we cannot
do in a naturalistic setting. While laboratory studies can generate useful qualitative insights, they
typically focus on quantitative research based on hypotheses because the controlled setting allows us
to manage external variables that may otherwise influence the results. This is the power of laboratory studies relative to naturalistic studies, surveys, or Web analytics.

    The specific major strength of a laboratory study lies in the investigation of dependent vari-
ables. In a laboratory study, we can design a setting to make changes to one or more independent

variables in order to investigate the effect on a dependent variable, while controlling for all the other variables (i.e., control variables). In such a setting, only the variable of interest affects the outcome.

There are multiple ways of designing a laboratory study. As such, laboratory studies can be very nuanced, and a review of laboratory studies is a lecture in itself. For more detailed examinations of laboratory studies and experiments, I refer the interested reader to the Controlled Standards of Reporting Trials (CONSORT) (www.consort-statement.org) that assist in the design of laboratory experiments, specifically randomized controlled trials. CONSORT provides a 22-item checklist and a flow diagram for conducting such studies. The checklist items focus on the study's design, analysis, and interpretation. The flow diagram illustrates the progress of participants through a laboratory study. Together, these tools aid in understanding the design and running of the study, the analysis of the collected data, and the interpretation of the results.

The Common Industry Format (CIF) is an American National Standards Institute (ANSI) approved standard for reporting the results of usability studies. The National Institute of Standards and Technology (NIST) developed this criterion to assist in designing and reporting the results of usability studies targeted specifically for Websites.

One good way to learn about laboratory studies is to read what others have done. What are some questions that one should ask when designing (or assessing) a laboratory study? One methodology for accessing laboratory studies is the Centre for Allied Heath Evidence (CAHE) Critical Appraisal Tools (CATs) (http://www.unisa.edu.au/cahe/CAHECATS/). The aim of the approach is to identify possible methodological flaws in the design phase or in the reporting. With the use of such a questionnaire, we can design better experiments and make informed decisions about the quality of research evidence. The assessments presented below are based on the CriSTAL Checklist for appraising a user study (http://www.shef.ac.uk/scharr/eblib/use.htm).

Does the study address a clearly focused issue? Essentially, the research aims should drive the study. The issue can deal with the population (user group) studied, the intervention (service or facility) provided, or the system. The laboratory study design must clearly identify the issue in a usable manner and explain how the outcomes (quantifiable or qualitative) are measured.

Is a good case made for the approach that the authors have taken? In designing a user study, researchers can choose from an extensive array of methods. One way to assess a study, then, is to review the selection of methodology (e.g., regression, ANOVA Analysis Of Variance, factor analysis, etc.) and design setup (e.g., within or between groups). The method and setup should relate directly to the research questions or objectives, which are tied to the research aim. A good study will clearly identify the problem and provide justification for the questions or objectives. The methodology must be appropriate to the research questions or objectives.

Were the methods used in selecting the users appropriate and clearly described? There are several aspects of recruiting participants for any laboratory study, including:

- **Type of sample:** This addresses how the participants are recruited. Most studies are convenience samples (i.e., you use who you can get). In academia, these are usually students, and the participants self-select into the study. Randomly selected sample are generally preferred; however, many times a convenience sample is not a critical shortcoming if the population demographics are not essential to the research questions.
- **Size of sample:** Sample size is an important aspect of user studies and it must be managed carefully. A sample size calculation (i.e., how many participants do you need to represent the population you are studying) can determine the appropriate size needed to make the sample representative of the population (i.e., does the sample represent targeted users). Representativeness matters for quantitative analysis. For usability studies, this is not so important. Generally, the demographics of the sample (e.g., age, sex, staff grade, location) must accurately reflect the demographics of the total population. Any motivation for the participants (i.e., money, course credit, etc.) must be acknowledged.
- **Was the data collection instrument/method reliable?** Any questionnaire, survey form, or interview schedule should be pilot tested before its use in the laboratory study. When adapting an instrument used in previous research, the case must be made for its appropriate use.

What was the response rate and how representative were respondents of the population under study? Whether using a convenience or random sample, researchers must ensure that no subgroups were either over-represented or under-represented. When using convenience samples for Web laboratory studies, sex is many times an issue.

Are the results complete and have they been analyzed in an easily interpretable way? Just as there are several choices in methodologies, there are also several choices concerning methods of analysis and in how to present these results. Regardless, the variables must be defined and identified.

Are there any limitations in the methodology (that might have influenced results) identified and discussed? No matter how well one designs a study, selects a sample, executes a methodology, and analyzes the data, there are always limitations. After the study is completed, reflect on how the study might be better implemented next time.

Are the conclusions based on an honest and objective interpretation of the results? Sometimes, a study does not tell you what you want or expect. That is actually good but frustrating. However, one must base the conclusions clearly on the findings from the study's data.

Just as log analysis and surveys have limitations, laboratory studies also have limitations that we must consider [94]. The basic assumption underlying laboratory studies and experiments is that we can extrapolate the results to the *real world*. However, when people are involved, this is a dicey assumption, and the validity of results from laboratory studies to contexts outside the laboratory is not flawless.

Some of the possible issues that can arise are laboratory effects (i.e., the context of the laboratory study is not a naturalistic setting), anonymity issues (i.e., the participants know they are being observed), context (i.e., regardless of the study design, there are aspects beyond the control of the researcher), and biased sample (i.e., regardless of the sampling method, there are biases created by participants self selection into the study and by the fact that they do not represent the portion of the population that never participates). By using Web analytics in conjunction with laboratory studies, we can address many of these shortcomings.

8.3   CONCLUSION

Web analytics via log data are an excellent means for recording the behaviors of system users and the responses of those systems. Because they focus on behavioral data only, however, transaction logs are ineffective as a method of understanding the underlying motivations, affective characteristics, cognitive factors, and contextual aspects that influence those behaviors. Used in conjunction with Web logs, surveys and laboratory studies can be effective methods for investigating these aspects. The combined methodological approaches can provide a richer picture of the phenomenon under investigation.

In this section, we have reviewed a 10-step procedure for conducting survey research, with explanatory notes on each step. We then discussed the design of a survey instrument, with examples of the various types of questions, and then discussed aspects of designing a laboratory study, providing some key questions that can help us in planning and completing a laboratory study.

CHAPTER 9

Search Log Analytics

A special case of Web analytics is analyzing data from search logs. Exploiting the data stored in search logs of Web search engines, Intranets, and Websites can provide important insights into understanding the information searching tactics of online users. This understanding can inform information system design, interface development, and information architecture construction for content collections.

This section presents a review of and foundation for conducting Web SLA [64, 66]. A basic understanding of search engines and searching behavior is assumed [for a review, see Ref. 93]. SLA methodology consists of three stages (i.e., *collection*, *preparation*, and *analysis)*, and those stages are presented in detail with discussions of the goals, metrics, and processes at each stage. Following this, the critical terms in TLA for Web searching are defined and suggestions are provided on ways to leverage the strengths and address the limitations of TLA for Web searching research.

9.1   INTRODUCTION

Information searching researchers have employed search logs for analyzing a variety of Web information systems [34, 73, 78, 151]. Web search engine companies use search logs (also referred to as transaction logs) to investigate searching trends and effects of system improvements (cf. Google at http://www.google.com/press/zeitgeist.html or Yahoo! at http://buzz.yahoo.com/buzz_log/?fr=fp-buzz-morebuzz). Search logs are an unobtrusive method of collecting significant amounts of searching data on a sizable number of system users. There are several researchers who have employed the SLA methodology to study Web searching. Romano et al. [128] present a methodology for general qualitative analysis of transaction log data. Wang et al. [151] and Spink and Jansen [140] also present explanations of approaches to TLA.

Generally, there are limited published works concerning how to employ search logs to support the study of Web searching, the use of Web search engines, Intranet searching, or other Web searching applications. Yet, SLA is helpful for studying Web searching on Websites and Web search engines.

9.2   REVIEW OF SEARCH ANALYTICS

9.2.1   What Is a Search Log?

Not surprisingly, a search log is a file (i.e., log) of the communications (i.e., transactions) between a system and the users of that system. Rice and Borgman [126] present transaction logs as a data collection method that automatically captures the type, content, or time of transactions made by a person from a terminal connected to that system. As noted previously, Peters [117] views transaction logs as electronically recorded interactions between on-line information retrieval systems and the persons who search for the information found in those systems.

For Web searching, a search log is *an electronic record of interactions that have occurred during a searching episode between a Web search engine and users searching for information on that Web search engine*. A Web search engine may be a general-purpose search engine, a niche search engine, a searching application on a single Website, or variations on these broad classifications. A searching episode is a period of user interaction with a search engine that may be composed of one or more sessions. The users may be humans or computer programs acting on behalf of humans. Interactions are the communication exchanges that occur between users and the system, and either the user or the system may initiate elements of these exchanges.

9.2.2   How Are These Interactions Collected?

The process of recording the data in the search log is relatively straightforward. Web servers record and store the interactions between searchers (i.e., actually Web browsers on a particular computer) and search engines in a log file (i.e., the search log) on the server using a software application. Thus, most search logs are server-side recordings of interactions. Major Web search engines execute millions of these interactions per day. The server software application can record various types of data and interactions depending on the file format that the server software supports.

As mentioned earlier, typical transaction log formats are access log, referrer log, and extended log. The W3C (http://www.w3.org/TR/WD-logfile.html) is one organizational body that defines transaction log formats. However, search logs are a special type of transaction log file. This search log format has most in common with the extended file format, which contains data such as the client computer's IP address, user query, search engine access time, and referrer site, among other fields.

9.2.3   Why Collect This Data?

Collecting the data enables us to analyze it in order to obtain beneficial information. Web analytics, as we have discussed, can focus on many interaction issues and research questions [36], but it typically addresses either issues of system performance, information structure, or user interactions. Blecic et al. [18] define TLA as the detailed and systematic examination of each search command or query by a user and the following database result or output. Phippen et al. [120] and Spink and Jansen [140] also provide comparable definitions of TLA.

For Web searching research, we focus on a subset of Web analytics, namely SLA. Web analytics is useful for analyzing the browsing or navigation patterns within a Website, while SLA is concerned exclusively with searching behaviors. SLA is defined as *the use of data collected in a search log to investigate particular research questions concerning interactions among Web users, the Web search engine, or the Web content during searching episodes.* Within this interaction context, we can exploit the data in search logs to discern attributes of the search process, such as the searcher's actions on the system, the system responses, or the evaluation of results by the searcher.

The goal of SLA is to gain a clearer understanding of the interactions among searcher, content and system or the interactions between two of these structural elements, based on whatever research questions drive the study. Employing SLA allows us to achieve some stated objective, such as improved system design, advanced searching assistance, or better understanding of some user information searching behavior.

9.2.4   What Is the Foundation of Search Log Analysis?

SLA lends itself to a grounded theory approach [44]. This approach emphasizes a systematic discovery of theory from data using methods of comparison and sampling. The resulting theories or models are grounded in observations of the real world, rather than being abstractly generated. Therefore, grounded theory is an inductive approach to theory or model development, rather than the deductive alternative [26].

In other words, when using SLA as a methodology, we examine the characteristics of searching episodes in order to isolate trends and identify typical interactions between searchers and the system. Interaction has several meanings in information searching, addressing a variety of transactions including query submission, query modification, results list viewing, and use of information objects (e.g., Web page, pdf file, and video).

For the purposes of SLA, we consider interactions the physical expressions of communication exchanges between the searcher and the searching system. For example, a searcher may submit a query (i.e., an interaction). The system may respond with a results page (i.e., an interaction). The searcher may click on a uniform resource locator (URL) in the results listing (i.e., an interaction). Therefore, for SLA, interaction is a mechanical expression of underlying information needs or motivations.

9.2.5   How Is Search Log Analysis Used?

Researchers and practitioners have used SLA (usually referred to as TLA in these studies) to evaluate library systems, traditional information retrieval (IR) systems, and more recently Web systems. Search logs have been used for many types of analysis; in this review, we focus on those studies that centered on or about searching. Peters [117] provides a review of TLA in library and experimental IR systems. Some progress has been made in TLA methods since Peters' summary [117] in terms of collection and ability to analyze data. Jansen and Pooch [69] report on a variety of studies employing SLA for the study of Web search engines and searching on Websites. Jansen and Spink [71] provide a comprehensive review of Web searching SLA studies. Other review articles include Kinsella and Bryant [87] and Fourie [42].

Several researchers have viewed SLA as a high-level designed process, including Copper [31]. Other researchers, such as Hancock-Beaulieu et al. [49], Griffiths et al. [47], Bains [12], Hargittai [51], and Yuan and Meadows [163], have advocated using SLA in conjunction with other research methodologies or data collection. Alternatives for other data collection include surveys and laboratory studies.

9.2.6   How to Conduct Search Log Analysis?

Despite the abundant literature on SLA, there are few published manuscripts on how actually to conduct it, especially with respect to SLA for Web searching. While some works provide fairly comprehensive descriptions of the methods employed, including Cooper [31], Nicholas et al. [108], Wang et al. [151], and Spink and Jansen [140], none presents a process or procedure for actually conducting TLA in sufficient detail to replicate the method. We will attempt to address this shortcoming, building on work presented in Ref. [66].

9.3   SEARCH LOG ANALYSIS PROCESS

Naturally, research questions need to be articulated in order to determine what data needs to be collected. However, search logs are typically of standard formats due to previously developed software applications. Given the interactions between users and Web browsers, which are the interfaces to Web search engines, the type of data that one can collect is standard. Therefore, the SLA methodology discussed here is applicable to a wide range of studies.

SLA involves three major stages, namely:

- **Data collection**: the process of collecting the interaction data for a given period in a transaction log;
- **Preparation**: the process of cleaning and preparing the transaction log data for analysis; and
- **Analysis**: the process of analyzing the prepared data.

9.3.1   Data Collection

The research questions define what information one must collect in a search log. Transaction logs provide a good balance between collecting a robust set of data and unobtrusively collecting that data [102]. If we are conducting a naturalistic study (i.e., outside of the laboratory) on a real system (i.e., a system used by actual searchers), then the method of data monitoring and collecting should not interfere with the information searching process. In addition to the loss of potential customers, a data collection method that interferes with the information searching process may unintentionally alter that process. For these reasons, and others, collecting data from real users pursuing needed information while interacting with real systems on the Web necessarily affects the type of data realistically obtainable.

9.3.2   Fields in a Standard Search Log

Table 9.1 provides a sample of a standard search log format collected by a Web search engine.

The fields are common in standard Web search engine logs, although some systems may log additional fields. A common additional field is a cookie identification code that facilitates identifying individual searchers using a common computer. A cookie is a text message given by a Web server to a Web browser and is stored on the client machine.

In order to facilitate valid comparisons and contrasts with other analysis, a standard terminology and set of metrics [69] is advocated. This standardization will help address one of Kurth's critiques [91] concerning the communication of SLA results across studies. Others have also noted terminology as an issue in Web research [121]. The standard field labels and descriptors are presented below.

A *searching episode* is a series of searching interactions within a given temporal span by a single searcher. Each record, shown as a row in Table 9.1, is a *searching interaction*. The format of each *searching interaction* is:

- *User Identification:* the IP address of the client's computer. This is sometimes also an anonymous user code address assigned by the search engine server, which is our example in Table 9.1.
- *Date:* the date of the interaction as recorded by the search engine server.
- *The Time:* the time of the interaction as recorded by the search engine server.
- *Search URL:* the query terms as entered by the user.
- Web search engine server software normally records these fields. Other common fields include *Results Page* (a code representing a set of result abstracts and URLs returned by the search engine in response to a query),
- *Language* (the user preferred language of the retrieved Web pages),

TABLE 9.1: Web search engine search log snippet.

USER IDENTIFICATION	DATE	TIME	QUERY
ce00	25/Apr/2009	04:08:50	Sphagnum Moss Harvesting + New Jersey + Raking
38f0	25/Apr/2009	04:08:50	emailanywhere
fabc	25/Apr/2009	04:08:54	Tailpiece
5010	25/Apr/2009	04:08:54	1'personalities AND gender AND education'1
25/Apr/2009	**04:08:54**	**dmr panasonic**	
89bf2	25/Apr/2009	04:08:55	bawdy poems"
	"Web Analytics	**25/Apr/2009**	
397e0		**04:08:56**	**gay and happy**
a9560	25/Apr/2009	04:08:58	skin diagnostic
81343	25/Apr/2009	04:08:59	Pink Floyd cd label cover scans
3c5c	25/Apr/2009	04:09:00	freie stellen dangaard
9daf	25/Apr/2009	04:09:00	Moto.it
415	25/Apr/2009	04:09:00	Capablity Maturity Model VS.
c03	25/Apr/2009	04:09:01	ana cleonides paulo fontoura

Note: Items in boldface are intentional errors.

- *Source* (the federated content collection searched, also known as *vertical*), and
- *Page Viewed* (the URL that the searcher visited after entering the query and viewing the results page, which is also known as either *click-thru or click-through*).

9.3.3 Data Preparation

Once the data is collected, we are ready to prepare the data. For data preparation, the focus is on importing the search log data into a relational database (or other analysis software), assigning each record a primary key, cleaning the data (i.e., checking each field for bad data), and calculating standard interaction metrics that will serve as the basis for further analysis.

Figure 9.1 shows the Entity–Relation (ER) diagram for the relational database that will be used to store and analyze the data from our search log. The ER diagram will vary based on the specific analysis.

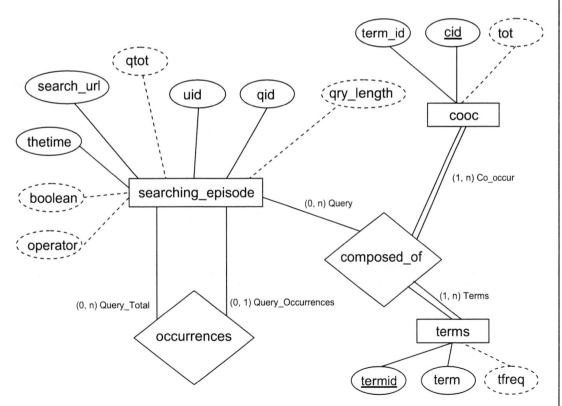

FIGURE 9.1: Web search log ER scheme diagram [66].

An ER diagram models the concepts and perceptions of the data and displays the conceptual schema for the database using standard ER notation. Table 9.2 presents the legend for the schema constructs.

Since search logs are in ASCII format, we can easily import the data into most relational databases. A key thing is to import the data in the same coding schema in which it was recorded

TABLE 9.2: Search log legend for ER schema constructs [66].

ENTITY NAME	CONSTRUCT
Searching_Episodes	a table containing the searching interactions
boolean	denotes if the query contains Boolean operators
operators	denotes if the query contains advanced query operators
q_length	query length in terms
qid	primary key for each record
qtot	number of results pages viewed
searcher_url	query terms as entered by the searcher
thetime	time of day as measured by the server
uid	user identification based on IP
Terms	table with terms and frequency
term_ID	term identification
term	term from the query set
tfreq	number of occurrences of term in the query set
Cooc	table term pairs and the number of occurrences of those pairs
term_ID	term identification
cid	the combined term identification for a pair of terms
tot	number of occurrences of the pair in the query set

(e.g., UTF-8, US-ASCII). Once imported, each record is assigned a unique identifier or primary key. Most modern databases can assign this automatically on importation, or we can assign it later using scripts.

9.3.4 Cleaning the Data

Once the search log data is in a suitable analysis software package, the focus shifts to cleaning the data. Records in search logs can contain corrupted data. These corrupted records have multiple causes, but they are mostly the result of errors in logging the data. In the example shown in Table 9.1, the errors are easy to spot (additionally, these records are rendered in boldface), but often a search log will number millions if not billions of records. Therefore, a visual inspection is not practical for error identification. From experience, one method of rapidly identifying most errors is to sort each field in sequence. Since the erroneous data will not fit the pattern of the other data in the field, these errors will usually appear at the top of, bottom of, or in groups in each sorted field. Standard database functions to sum and group key fields, such as time and IP address, will usually identify any further errors. We must remove all records with corrupted data from the transaction log database. Typically, the percentage of corrupted data is small relative to the overall database.

9.3.5 Parsing the Data

To demonstrate how to parse data, we will use the three fields of *The Time*, *User Identification*, and *Search URL*, common to all Web search logs to recreate the chronological series of actions in a searching episode. The Web query search logs usually contain queries from both human users and agents. Depending on the research objective, we may be interested in only individual human interactions, those from common user terminals, or those from agents. For the running example used here, we will consider the case of only having an interest in human searching episodes.

Given that there is no way to accurately identify human from non-human searchers [135, 146], most researchers using Web search log either ignore it [25] or assume some temporal or interaction cutoff [107, 135]. Using a cutoff of 101 queries, the subset of the search log is weighted to queries submitted primarily by human searchers in a non-common user terminal, but 101 queries is high enough not to introduce bias by too low of a cutoff threshold. The selection of 101 is arbitrary, and other researchers have used a wide variety of cutoffs. For our example, we will separate all sessions with fewer than 101 queries into an individual search log.

There are several methods to remove these large sessions. One can code a program to count the session lengths and then delete all sessions that have lengths over 100. For smaller log files (a few million or so records), it is just as easy to do with SQL queries. To do this, we must first remove records that do not contain queries. From experience, search logs may contain many such records (usually on the order of 35% to 40% of all records) as users go to Websites for purposes other than searching [72].

9.3.6   Normalizing Searching Episodes

When a searcher submits a query, then views a document, and returns to the search engine, the Web server typically logs this second visit with the identical user identification and query, but with a new time (i.e., the time of the second visit). This is beneficial information in determining how many of the retrieved *results pages* the searcher visited from the search engine, but unfortunately, it also skews the results in analyzing the query level of analysis. In order to normalize the searching episodes, we must first separate these result page requests from query submissions for each searching episode.

In SLA, researchers are often interested in terms and term usage, which can be an entire study in itself. In these cases, it is usually cleaner to generate separate tables that contain each term and their frequency of occurrence than to attempt combination tables. A term co-occurrence table that contains each term and its co-occurrence with other terms is also valuable for understanding the data. With a relational database, we can generate these tables using scripts. If using text-parsing languages, we can parse these terms and associated data during initial processing.

There are already several fields in our database, many of which can provide valuable information. From these items, we can calculate several metrics, some of which take a long time to compute for large datasets.

9.4   DATA ANALYSIS

This stage focuses on three levels of analysis. As we discuss these levels, we will step through the data analysis stage.

9.4.1   Analysis Levels

The three common levels of analysis for examining transaction logs are *term*, *query, and session.*

Term level analysis. The term level of analysis naturally uses the *term* as the basis for analysis. A *term* is a string of characters separated by some delimiter such as a space or some other separator. At this level of analysis, one focuses on measures such as *term occurrence*, which is the frequency that a particular term occurs in the transaction log. *Total terms* are the number of terms in the dataset. *Unique terms* are the terms that appear in the data regardless of the number of times they occur. *High Usage Terms* are those terms that occur most frequently in the dataset. *Term co-occurrence* measures the occurrence of term pairs within queries in the entire search log. We can also calculate degrees of association of term pairs using various statistical measures [cf. Refs. 131, 135, 151].

The mutual information formula measures term association and does not assume mutual independence of the terms within the pair. We calculate the mutual information statistic for all term pairs within the dataset. Many times, a relatively low frequency term pair may be strongly associated (i.e., if the two terms always occur together). The mutual information statistic identifies the strength of this association. The mutual information formula used in this research is

$$I(w_i, w_2) = \ln\frac{P(w_1, w_2)}{P(w_1)P(w_2)}$$

where $P(w_1)$, $P(w_2)$ are probabilities estimated by relative frequencies of the two words and $P(w_1, w_2)$ is the relative frequency of the word pair and order is not considered. Relative frequencies are observed frequencies (F) normalized by the number of the queries:

$$P(w_1) = \frac{F_1}{Q'}; P(w_1) = \frac{F_2}{Q'}; P(w_1, w_2) = \frac{F_2}{Q'}$$

Both the frequency of term occurrence and the frequency of term pairs are the occurrence of the term or term pair within the set of queries. However, since a one-term query cannot have a term pair, the set of queries for the frequency base differs. The number of queries for the terms is the number of non-duplicate queries in the dataset. The number of queries for term pairs is defined as:

$$Q' = \sum_{n}^{m}(2n-3)Q_n$$

where Q_n is the number of queries with n words ($n > 1$), and m is the maximum query length. So, queries of length one have no pairs. Queries of length two have one pair. Queries of length three have three possible pairs. Queries of length four have five possible pairs. This continues up to the queries of maximum length in the dataset. The formula for queries of term pairs (Q') account for this term pairing.

Query level analysis. The query level of analysis uses the query as the base metric. A *query* is defined as a string list of one or more terms submitted to a search engine. This is a mechanical definition as opposed to an information searching definition [88]. The first query by a particular searcher is the *initial query*. A subsequent query by the same searcher that is different from any of the searcher's other queries is a *modified query*. There can be several occurrences of different modified queries by a particular searcher. A subsequent query by the same searcher that is identical to one or more of the searcher's previous queries is an *identical query*.

In many Web search engine logs, when the searcher traverses to a new results page, this interaction is also logged as an *identical query*. In other logging systems, the application records the page rank. A results page is the list of results, either sponsored or organic (i.e., non-sponsored), returned by a Web search engine in response to a query. Using either *identical queries* or some results page field, we can analyze the result page viewing patterns of Web searchers.

Other measures are also observable at the query level of analysis. A *unique query* refers to a query that is different from all other queries in the transaction log, regardless of the searcher. A *repeat query* is a query that appears more than once within the dataset by two or more searchers.

Query complexity examines the query syntax, including the use of advanced searching techniques such as Boolean and other query operators. *Failure rate* is a measure of the deviation of queries from the published rules of the search engine. The use of query syntax that the particular IR system does not support, but may be common on other IR systems, is *carry over*.

Session level analysis. At the session level of analysis, we primarily examine the within-session interactions [48]. However, if the search log spans more than one day or assigns some temporal limit to interactions from a particular user, we could examine between-sessions interactions. A *session interaction* is any specific exchange between the searcher and the system (i.e., submitting a query, clicking a hyperlink, etc.). A *searching episod*e is defined as a series of interactions within a limited duration to address one or more information needs. This session duration is typically short, with Web researchers using between 5 and 120 minutes as a cutoff [cf. Refs. 54, 70, 107, 135]. Each choice of time has an impact on the results, of course. The searcher may be multitasking [106, 138] within a searching episode, or the episode may be an instance of the searcher engaged in successive searching [95, 110, 141]. This *session* definition is similar to the definition of a *unique visitor* used by commercial search engines and organizations to measure Website traffic. The number of queries per searcher is the *session length.*

Session duration is the total time the user spent interacting with the search engine, including the time spent viewing the first and subsequent Web documents, except the final document. Session duration can therefore be measured from the time the user submits the first query until the user departs the search engine for the last time (i.e., does not return). The viewing time of the final Web document is not available since the Web search engine server does not record the time stamp. Naturally, the time between visits from the Web document to the search engine may not have been entirely spent viewing the Web document, which is a limitation of the measure.

A *Web document* is the Web page referenced by the URL on the search engine's results page. A Web document may be text or multimedia and, if viewed hierarchically, may contain a nearly unlimited number of sub-Web documents. A Web document may also contain URLs linking to other Web documents. From the results page, a searcher may click on a URL, (i.e., visit) one or more results from the listings on the result page. This is *click through analysis* and measures the page viewing behavior of Web searchers. We measure *document viewing duration* as the time from when a searcher clicks on a URL on a results page to the time that searcher returns to the search engine.

Some researchers and practitioners refer to this type of analysis as *page view analysis. Click through analysis* is possible if the transaction log contains the appropriate data. There are many other factors one can examine, including query graphs [11].

9.4.2   Conducting the Data Analysis

The key to successful SLA is conducting the analysis with an organized approach. One method is to sequentially number and label the queries (or coded modules) to correspond to the order of execution and to their function, since many of these queries must be executed in a certain order to obtain valid results. Many relational database management systems provide mechanisms to add descriptive properties to the queries. These can provide further explanations of the query function

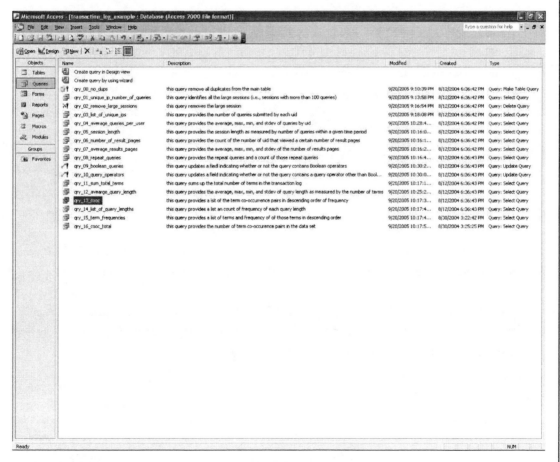

FIGURE 9.2: SLA numbered and descriptively labeled queries [66].

or relate these queries directly to research questions. Figure 9.2 illustrates the application of such an approach.

Figure 9.2 also shows each query in sequence and provides a descriptive tag describing that query's function.

SLA involves a series of standard analyses that are common to a wide variety of Web searching studies. Some of these analyses may directly address certain research questions, and others may be the basis for more in-depth research analysis.

One typical question is, "How many searchers have visited the search engine during this period?" This query will provide a list of unique searchers and the number of queries they have submitted during the period. We can modify this question and determine "How many searchers have visited the search engine on each day during this period." Naturally, a variety of statistical results can be determined using the previous queries. For example, we can determine the standard deviation of number of queries per searcher.

In addition to visits, we may want information about the session lengths (i.e., the number of queries within a session) for each searcher. Similarly, we may be curious about the number of searchers who viewed a certain number of results pages.

We can calculate various statistical results on results page viewing, such as the maximum number of result pages viewed and queries per day. An important aspect for system designers is results caching because we need to know the number of repeat queries submitted by the entire set of searchers during a given period in order to optimize our system's performance.

Some researchers are more interested in how searchers are interacting with a search engine, and for this purpose the use of Boolean operators is an important feature. Since most search engines offer other query syntax than just Boolean operators, we can also investigate the use of these other operators.

Counting the terms within the transaction log is another typical measurement. We certainly want to know about query length, the frequency of terms pairs, and the various term frequencies.

The results from this series of queries provide us with a wealth of information about our data (e.g., occurrences of session lengths, occurrences of query length, occurrences of repeat queries, most used terms, most used term pairs) and serves as the basis for further investigations (e.g., session complexity, query structure, query modifications, term relationships).

9.5   DISCUSSION

It is certainly important to understand both the strengths and limitations of SLA for Web searching. First concerning the strengths, SLA provides a method of collecting data from a great number of users. Given the current nature of the Web, search logs appears to be a reasonable and non-intrusive means of collecting user system interaction data during the Web information searching process from a large number of searchers. We can easily collect data on hundreds of thousands to millions of interactions, depending on the traffic of the Website.

Second, we can collect this data inexpensively. The costs are the software and storage. Third, the data collection is unobtrusive, so the interactions represent the unaltered behavior of searchers, assuming the data is from an operational searching site. Finally, search logs are, at present, the only method for obtaining significant amounts of search data within the complex environment that is the Web [37]. Of course, researchers can also undertake SLA from research sites or capture client-side data across multiple sites using a custom Web browser (for the purpose of data collection) that does not completely mimic the searcher's natural environment.

There are limitations with SLA, as with any methodology. First, certain types of data are not in the transaction log, individuals' identities being the most common example. An IP address typically represents the "user" in a search log. Since more than one person may use a computer, an IP address is an imprecise representation of the user. Search engines are overcoming this limitation somewhat by the use of cookies.

Second, there is no way to collect demographic data when using search logs in a naturalistic setting. This constraint is true of many non-intrusive naturalistic studies. However, there are several sources for demographic data on the Web population based on observational and survey data. From these data sources we may get reasonable estimations of needed demographic data. However, this demographic data is still not attributable to specific subpopulations.

Third, a search log does not record the reasons for the search, the searcher motivations, or other qualitative aspects of use. This is certainly a limitation. In the instances where one needs this data, one should use TLA in conjunction with other data collection methods. However, this invasiveness reduces the unobtrusiveness, which is an inherent advantage of search logs as a data collection method.

Fourth, the logged data may not be complete due to caching of server data on the client machine or proxy servers. This is an often-mentioned limitation. In reality, this is a relatively minor concern for Web search engine research due to the method with which most search engines dynamically produce their results pages. For example, a user accesses the page of results from a search engine using the *Back* button of a browser. This navigation accesses the results page via the cache on the client machine. The Web server will not record this action. However, if the user clicks on any URL on that results page, functions coded on the results page redirects the click first to the Web server, from which the Web server records the visit to the Website.

9.6   CONCLUSION

We presented a three-step methodology for conducting SLA, namely collecting, preparing, and analyzing. We then reviewed each step in detail, providing observations, guides, and lessons learned.

We also discussed the organization of the database at the ER-level, and we explained the table design for standard search engine transaction logs. This presentation of the methodology at a detailed level of granularity will serve as an excellent basis for novice or experienced search log researchers.

Search logs are powerful tools for collecting data on the interactions between users and systems. Using this data, SLA can provide significant insights into user–system interactions, and it complements other methods of analysis by overcoming the limitations inherent in those methods. By combining SLA with other data collection methods or other research results, we can improve the robustness of the analysis. Overall, SLA is a powerful tool for Web searching research, and the SLA process outlined here can be helpful in future Web searching research endeavors.

*   *   *   *

CHAPTER 10

Conclusion

This lecture presents an overview of the Web analytics process, with a focus on gaining insight and actionable outcomes from collecting and analyzing Internet data. The lecture first provides an overview of Web analytics, providing in essence, a condensed version of the entire lecture. The lecture then outlines the theoretical and methodological foundations of Web analytics in order to understand clearly the strengths and shortcomings of Web analytics as an approach. These foundational elements include the psychological basis in behaviorism and methodology underpinning of trace data as an empirical method. The lecture then presents a brief history of Web analytics from the original transaction log studies in the 1960s, through the information science investigations of library systems, to the focus on Websites, systems, and applications. The lecture then covers the various types of ongoing interaction data within the clickstream created using log files and page tagging for analytics of Website and search logs. The lecture then presents a Web analytic process to convert this basic data to meaningful KPIs to measure likely converts that are tailored to the organizational goals or potential opportunities. Supplementary data collection techniques are addressed, including surveys and laboratory studies. The lecture then discusses the strengths and shortcoming of Web analytics. The overall goal of this lecture is to provide implementable information and a methodology for understanding Web analytics in order to improve Web systems, increase customer satisfaction, and target revenue through effective analysis of user–Website interactions.

Returning to that online retail store selling the latest athletic shoe, Web analytics can tell us how potential customers find our online store, including those who are referred from other Websites and those from search engines. Web analytics provides us the methods to know, and our KPIs tell us why we should care. Our understanding of customer behavior provided by Web analytics gives us the tool to determine what it might mean if customers come to our Website and then immediately leave versus if the potential customer explores several pages and then leaves. We can leverage Web analytics techniques to glean value from this data. Web analytics allows us to focus on organization goals, including getting the customer through the entire shopping cart process. In sum, Web analytics is the strategic tool to make our hypothetical online store successful by understanding why potential customers behave as they do and what that behavior means.

■　■　■　■

CHAPTER 11

Key Terms

- **Abandonment rate**: key performance indicator that measures the percentage of visitors who got to that point on the site but decided not to perform the target action.
- **Alignment-centric performance management**: method of defining a site's business goals by choosing only a few key performance indicators.
- **Average order value**: key performance indicator that measures the total revenue to the total number of orders.
- **Average time on site**: see *visit length*.
- **Behavior**: essential construct of the behaviorism paradigm. At its most basic, a behavior is an observable activity of a person, animal, team, organization, or system. Like many basic constructs, behavior is an overloaded term because it also refers to the aggregate set of responses to both internal and external stimuli. Therefore, behaviors address a spectrum of actions. Because of the many associations with the term, it is difficult to characterize it without specifying a context in which it takes place to provide meaning.
- **Behaviorism**: research approach that emphasizes the outward behavioral aspects of thought. For transaction log analysis, we take a more open view of behaviorism. In this more encompassing view, behaviorism emphasizes the observed behaviors without discounting the inner aspects that may accompany these outward behaviors.
- **Checkout conversion rate**: key performance indicator that measures the percent of total visitors who begin the checkout process.
- **Commerce Website**: a type of Website where the goal is to get visitors to purchase goods or services directly from the site.
- **Committed visitor index**: key performance indicator that measures the percentage of visitors that view more than one page or spend more than 1 minute on a site (these measurements should be adjusted according to site type).
- **Content/media Website**: a type of Website focused on advertising.
- **Conversion rate**: key performance indicator that measures the percentage of total visitors to a Website that perform a specific action.
- **Cost per lead (CPL)**: key performance indicator that measures the ratio of marketing expenses to total leads and shows how much it costs a company to generate a lead.

- **Customer loyalty**: key performance indicator that measures the ratio of new to existing customers.
- **Customer satisfaction metrics**: key performance indicator that measures how the users rate their experiences on a site.
- **Demographics and system statistics**: a metric that measures the physical location and information of the system used to access the Website.
- **Depth of visit**: key performance indicator that measures the ratio between page views and visitors.
- **Electronic survey**: method of data collection in which a computer plays a major role in both the delivery of a survey to potential respondents and the collection of survey data from actual respondents.
- **Ethogram**: index of the behavioral patterns of a unit. An ethogram details the different forms of behavior that an actor displays. In most cases, it is desirable to create an ethogram in which the categories of behavior are objective, discrete, and not overlapping with each other. The definitions of each behavior should be clear, detailed, and distinguishable from each other. Ethograms can be as specific or general as the study or field warrants.
- **Interactions:** physical expressions of communication exchanges between the searcher and the system.
- **Internal search**: a metric that measures information on keywords and results pages viewed using a search engine embedded in the Website.
- **Key performance indicator (KPI)**: a combination of metrics tied to a business strategy.
- **Lead generation Website**: Website used to obtain user contact information in order to inform them of a company's new products and developments and to gather data for market research.
- **Log file**: log kept by a Web server of information about requests made to the Website including (but not limited to) visitor IP address, date and time of the request, request page, referrer, and information on the visitor's Web browser and operating system.
- **Log file analysis**: method of gathering metrics that uses information gathered from a log file to gather Website statistics.
- **Metrics**: statistical data collected from a Website such as number of unique visitors, most popular pages, etc.
- **New visitor**: a user who is accessing a Website for the first time.
- **New visitor percentage**: key performance indicator that measures the ratio of new visitors to unique visitors.

- **Online business performance management (OBPM)**: method of defining a site's business goals that emphasizes the integration of business tools and Web analytics to make better decisions quickly in an ever-changing online environment.
- **Order conversion rate**: key performance indicator that measures the percentage of total visitors who place an order on a Website.
- **Page depth**: key performance indicator that measures the ratio of page views for a specific page and the number of unique visitors to that page.
- **Page tagging**: method of gathering metrics that uses an invisible image to detect when a page has been successfully loaded and then uses JavaScript to send information about the page and the visitor back to a remote server.
- **Prospect rate**: key performance indicator that measures the percentage of visitors who get to the point in a site where they can perform the target action (even if they do not actually complete it).
- **Referrers and keyword analysis**: a metric that measures which sites have directed traffic to the Website and which keywords visitors are using to find the Website.
- **Repeat visitor**: a user who has been to a Website before and is now returning.
- **Returning visitors**: key performance indicator that measures the ratio of unique visitors to total visits.
- **Search engine referrals**: key performance indicator that measures the ratio of referrals to a site from specific search engines to the industry average.
- **Search log analysis (SLA)**: use of data collected in a search log to investigate particular research questions concerning interactions among Web users, the Web search engine, or the Web content during searching episodes.
- **Search log analysis (SLA) process**: three stage process of collection, preparation, and analysis.
- **Search log**: electronic record of interactions that have occurred during a searching episode between a Web search engine and users searching for information on that Web search engine.
- **Single access ratio**: key performance indicator that measures the ratio of total single access pages (or pages where the visitor enters the site and exits immediately from the same page) to total entry pages.
- **Stickiness**: key performance indicator that measures how many people arrive at a homepage and proceed to traverse the rest of the site.
- **Support/self-service Website**: a type of Website that focuses on helping users find specialized answers for their particular problems.

- **Survey instruments:** a data collection procedure used in a variety of research designs.
- **Survey research:** a method for gathering information by directly asking respondents about some aspect of themselves, others, objects, or their environment.
- **Top pages**: a metric that measures the pages in a Website that receive the most traffic.
- **Total bounce rate**: key performance indicator that measures the percentage of visitors who scan the site and then leave.
- **Trace data**: measures that offer a sharp contrast to directly collected data. The greatest strength of trace data is that it is unobtrusive. The collection of the data does not interfere with the natural flow of behavior and events in the given context. Since the data is not directly collected, there is no observer present in the situation where the behaviors occur to affect the participants' actions. Trace data is unique; as unobtrusive and nonreactive data, it can make a very valuable research course of action. In the past, trace data was often time consuming to gather and process, making such data costly. With the advent of transaction logging software, trace data for the studying of behaviors of users and systems has become popular.
- **Traffic concentration**: key performance indicator that measures the ratio of number of visitors to a certain area in a Website to total visitors.
- **Transaction log**: electronic record of interactions that have occurred between a system and users of that system. These log files can come from a variety of computers and systems (Websites, OPAC, user computers, blogs, listserv, online newspapers, etc.), basically any application that can record the user–system–information interactions.
- **Transaction log analysis (TLA)**: broad categorization of methods that covers several sub-categorizations, including Web log analysis (i.e., analysis of Web system logs), blog analysis, and search log analysis (analysis of search engine logs).
- **Unique visit**: one visit to a Website (regardless of if the user has previously visited the site); an alternative to unique visitors.
- **Unique visitor**: a specific user who accesses a Website.
- **Unobtrusive methods**: research practices that do not require the researcher to intrude in the context of the actors. Unobtrusive methods do not involve direct elicitation of data from the research participants or actors. This approach is in contrast to obtrusive methods such as laboratory experiments and surveys that require researchers to physically interject themselves into the environment being studied.
- **Visit length**: a metric that measures total amount of time a visitor spends on the Website.
- **Visit value**: key performance indicator that measures the total number of visits to total revenue.

- **Visitor path**: a metric that measures the route a visitor uses to navigate through the Website.
- **Visitor type**: a metric that measures users who access a Website. Each user who visits the Website is a unique user. If it is a user's first time to the Website, that visitor is a new visitor, and if it is not the user's first time, that visitor is a repeat visitor.
- **Web analytics**: the measurement of visitor behavior on a Website.
- **Web analytics:** the measurement, collection, analysis, and reporting of Internet data for the purposes of understanding and optimizing Web usage (http://www.webanalytics-association.org/).

*   *   *   *

CHAPTER 12

Blogs for Further Reading

Listed below are several practitioner blogs that offer current and insightful analysis on Web analytics.

- **Analytics Notes by Jacques Warren**, http://www.waomarketing.com/blog
- **Occam's Razor by Avinash Kaushik**, http://www.kaushik.net/avinash/
- **SemAngel: Web Analytics and SEM Analytics by Gary Angel**, http://semphonic.blogs .com/
- **Web Analytics Demystified by Eric Peterson**, http://blog.webanalyticsdemystified.com/ weblog/
- **Web Analytics Articles by Jim Sterne**, http://www.emetrics.org/articlesbysterne.php

References

[1]   S. F. Abdinnour-Helm, B. S. Chaparro, and S. M. Farmer, "Using the End-User Comput- ing Satisfaction (EUCS) Instrument to Measure Satisfaction with a Web Site," *Decision Sciences,* vol. 36, pp. 341–364, 2005.

[2]   G. Abdulla, B. Liu, and E. Fox, "Searching the World-Wide Web: Implications from Studying Different User Behavior," in *the World Conference of the World Wide Web, Internet, and Intranet*, Orlando, FL, 1998, pp. 1–8.

[3]   S. E. Aldrich, "The Other Search: Making the Most of Site Search to Optimize the Total Customer Experience, " 6 June 2006, retrieved 14 May 2009 from http://www.docuticker .com/?p=5508.

[4]   H. Aljifri and D. S. Navarro, "Search engines and privacy," *Computers & Security,* vol. 23, pp. 379–388, 2004.

[5]   S. Ansari, R. Kohavi, L. Mason, and Z. Zheng, "Integrating E-Commerce and Data Mining: Architecture and Challenges," *IEEE International Conference on Data Mining,* pp. 27–34, 2001.

[6]   A. Avinash, "Bounce Rate: Sexiest Web Metric Ever?," 26 June 2007, retrieved 15 May 2009 from http://www.mpdailyfix.com/2007/06/bounce_rate_sexiest_web_metric.html.

[7]   R. Baeza-Yates, L. Calderón-Benavides, and C. González, "The Intention Behind Web Queries," in *String Processing and Information Retrieval (SPIRE 2006)*, Glasgow, Scotland, 2006, pp. 98–109.

[8]   R. Baeza-Yates and C. Castillo, "Relating Web Characteristics" [in Spanish], October 2000, retrieved 15 July 2002, from http://www.todocl.cl/stats/rbaeza.pdf.

[9]   R. Baeza-Yates and C. Castillo, "Relating Web Structure and User Search Behavior," in *10th World Wide Web Conference*, Hong Kong, China, 2001, pp. 1–2.

[10] R. Baeza-Yates, A. Gionis, F. Junqueira, V. Murdock, V. Plachouras, and F. Silvestri, "The Impact of Caching on Search Engines," in *30th annual international ACM SIGIR conference on Research and development in information retrieval*, Amsterdam, The Netherlands, 2007, pp. 183–190.

[11] R. Baeza-Yates and A. Tiberi, "The Anatomy of a Large Query Graph," *Journal of Physics A: Mathematical and Theoretical*, vol. 41, pp. 1–13, 2008.

[12] S. Bains, "End-User Searching Behavior: Considering Methodologies," *The Katharine Sharp Review*, vol. 1, http://www.lis.uiuc.edu/review/winter1997/bains.html, 1997.

[13] J. Bar-Ilan, "The Use of Web Search Engines in Information Science Research," in *Annual Review of Information Science and Technology*, vol. 33, B. Cronin, Ed. Medford, NY, USA: Information Today, 2004, pp. 231–288.

[14] J. D. Becher, "Why Metrics-Centric Performance Management Solutions Fall Short," in *Information Management Magazine*, vol. March, 2005.

[15] S. M. Beitzel, E. C. Jensen, A. Chowdhury, D. Grossman, and O. Frieder, "Hourly Analysis of a Very Large Topically Categorized Web Query Log," in *the 27th annual international conference on research and development in information retrieval*, Sheffield, UK, 2004, pp. 321–328.

[16] S. M. Beitzel, E. C. Jensen, D. D. Lewis, A. Chowdhury, and O. Frieder, "Automatic classification of Web queries using very large unlabeled query logs," *ACM Transactions on Information Systems*, vol. 25, no. 9, 2007.

[17] M. Belkin, "15 Reasons Why All Unique Visitors Are Not Created Equal," 8 April 2006, retrieved 15 May 2009 from http://www.omniture.com/blog/node/16.

[18] D. Blecic, N. S. Bangalore, J. L. Dorsch, C. L. Henderson, M. H. Koenig, and A. C. Weller, "Using transaction log analysis to improve OPAC retrieval results," *College & Research Libraries*, vol. 59, pp. 39–50, 1998.

[19] D. L. Booth and B. J. Jansen, "A Review of Methodologies for Analyzing Websites," in *Handbook of Research on Web Log Analysis*, B. J. Jansen, A. Spink, and I. Taksa, Eds. Hershey, PA: IGI, 2008, pp. 143–164.

[20] B. R. Boyce, C. T. Meadow, and D. H. Kraft, *Measurement in Information Science*. Orlando, FL: Academic Press Inc., 1994.

[21] N. Brooks, "The Atlas Rank Report I: How Search Engine Rank Impacts Traffic," July 2004, retrieved 1 August 2004 from http://www.atlasdmt.com/media/pdfs/insights/RankReport.pdf.

[22] N. Brooks, "The Atlas Rank Report II: How Search Engine Rank Impacts Conversions," October 2004, retrieved 15 January 2005 from http://www.atlasonepoint.com/pdf/Atlas RankReportPart2.pdf.

[23] J. Burby, "Build a Solid Foundation With Key Performance Indicators, Part 1: Lead-Generation Sites," 20 July 2004, retrieved 30 May 2009 from http://www.clickz.com/showPage.html?page=3382981.

[24] J. Burby and S. Atchison, *Actionable Web Analytics: Using Data to Make Smart Business Decisions*. Indianapolis, IN: Wiley, 2007.

[25] F. Cacheda and Á. Viña, "Experiences Retrieving Information in the World Wide Web," in *6th IEEE Symposium on Computers and Communications*, Hammamet, Tunisia, 2001, pp. 72–79.

[26] K. Chamberlain, "What is Grounded Theory?," 6 November 1995, retrieved 17 September 2005 from http://kerlins.net/bobbi/research/qualresearch/bibliography/gt.html.

[27] H.-M. Chen and M. D. Cooper, "Stochastic modeling of usage patterns in a web-based information system," *Journal of the American Society for Information Science and Technology*, vol. 53, pp. 536–548, 2002.

[28] H.-M. Chen and M. D. Cooper, "Using clustering techniques to detect usage patterns in a Web-based information system," *Journal of the American Society for Information Science and Technology*, vol. 52, pp. 888–904, 2001.

[29] C. Choo, B. Detlor, and D. Turnbull, "A Behavioral Model of Information Seeking on the Web: Preliminary Results of a Study of How Managers and IT Specialists Use the Web," in *61st Annual Meeting of the American Society for Information Science*, Pittsburgh, PA, 1998, pp. 290–302.

[30] A. Chowdhury and I. Soboroff, "Automatic Evaluation of World Wide Web Search Services," in *25th Annual International ACM SIGIR Conference on Research and Development in Information Retrieval*, Tampere, Finland, 2002, pp. 421–422.

[31] M. D. Cooper, "Design considerations in instrumenting and monitoring Web-based information retrieval systems," *Journal of the American Society for Information Science*, vol. 49, pp. 903–919, 1998.

[32] V. Cothey, "A longitudinal study of World Wide Web users' information searching behavior," *Journal of the American Society for Information Science and Technology*, vol. 53, pp. 67–78, 2002.

[33] M. Couper, "Web surveys: A review of issues and approaches," *Public Opinion Quarterly*, vol. 64, pp. 464–494, 2000.

[34] W. B. Croft, R. Cook, and D. Wilder, "Providing Government Information on the Internet: Experiences with THOMAS," in *the Digital Libraries Conference*, Austin, TX, 1995, pp. 19–24.

[35] D. A. Dillman, *Mail and Telephone Surveys*. New York: John Wiley & Sons, 1978.

[36] M. C. Drott, "Using Web Server Logs to Improve Site Design," in *the 16th Annual International Conference on Computer Documentation*, Quebec, Canada, 1998, pp. 43–50.

[37] S. T. Dumais, "Web Experiments and Test Collections," 7–11 May 2002, retrieved 20 April 2003 from http://www2002.org/presentations/dumais.pdf.

[38] N. Eiron and K. McCurley, "Analysis of Anchor Text for Web Search," in *the 26th Annual International ACM SIGIR Conference on Research and Development in Information Retrieval*, Toronto, Canada, 2003, pp. 459–460.

[39] eVision, "Websites that convert visitors into customers: Improving the ability of your Website to convert visitors into inquiries, leads, and new business," 27 September 2007, retrieved 15 May 2009 from http://www.evisionsem.com/marketing/webanalytics.htm.

[40] A. Fink, *The Survey Handbook (Vol. 1)*. Thousands Oaks, CA: Sage Publications, 1995.

[41] FoundPages, "Increasing Conversion Rates," 25 October 2007, retrieved 15 May 2009 from http://www.foundpages.com/calgary-internet-marketing/search-conversion.html.

[42] I. Fourie, "A Review of Web Information-seeking/searching studies (2000–2002): Implications for Research in the South African Context," in *Progress in Library and Information Science in Southern Africa: 2d Biannial DISSAnet Conference*, Pretoria, South Africa, 2002, pp. 49–75.

[43] F. J. Fowler, *Improving Survey Questions: Design and Evaluation (Vol. 38)*. Thousand Oaks, CA: Sage Publications, 1995.

[44] B. Glaser and A. Strauss, *The Discovery of Grounded Theory: Strategies for Qualitative Research*. Chicago, IL: Aldine Publishing, 1967.

[45] A. M. Graziano and M. L. Raulin, *Research Methods: A Process of Inquiry*, 5th ed. Boston: Pearson, 2004.

[46] M. Greenfield, "Use Web Analytics to Improve Profits for New Year: Focus on Four Key Statistics," 1 January 2006, retrieved 14 May 2009 from http://www.practicalecommerce.com/articles/132/Use-Web-Analytics-to-Improve-Profits-for-New-Year/.

[47] J. R. Griffiths, R. J. Hartley, and J. P. Willson, "An improved method of studying user-system interaction by combining transaction log analysis and protocol analysis," *Information Research*, vol. 7, http://InformationR.net/ir/7-4/paper139.html, 2002.

[48] M. Hancock-Beaulieu, "Interaction in information searching and retrieval," *Journal of Documentation*, vol. 56, pp. 431–439, 2000.

[49] M. Hancock-Beaulieu, S. Robertson, and C. Nielsen, "Evaluation of online catalogues: an assessment of methods (BL Research Paper 78)," The British Library Research and Development Department, London, 1990.

[50] M. H. Hansen and E. Shriver, "Using Navigation Data to Improve IR Functions in the Context of Web Search," in the *Tenth International Conference on Information and Knowledge Management*, Atlanta, GA, 2001, pp. 135–142.

[51] E. Hargittai, "Beyond logs and surveys: In-depth measures of people's web use skills," *Journal of the American Society for Information Science and Technology*, vol. 53, pp. 1239–1244, 2002.

[52] E. Hargittai, "Classifying and coding online actions," *Social Science Computer Review*, vol. 22, pp. 210–227, 2004.

[53] K. Hawkey, "Privacy Issues Associated with Web Logging Data," in *Handbook of Research on Web Log Analysis*, B. J. Jansen, A. Spink, and I. Taksa, Eds. Hershey, PA: IGI, 2008, pp. 80–99.

[54] D. He, A. Göker, and D. J. Harper, "Combining evidence for automatic Web session identification," *Information Processing & Management*, vol. 38, pp. 727–742, September 2002.

[55] D. Hilbert and D. Redmiles, "Agents for Collecting Application Usage Data Over the Internet," in *Second International Conference on Autonomous Agents (Agents '98)*, Minneapolis/St. Paul, MN, 1998, pp. 149–156.

[56] D. M. Hilbert and D. F. Redmiles, "Extracting usability information from user interface events," *ACM Computing Surveys*, vol. 32, pp. 384–421, 2000.

[57] C. Hölscher and G. Strube, "Web search behavior of Internet experts and newbies," *International Journal of Computer and Telecommunications Networking*, vol. 33, pp. 337–346, 2000.

[58] O. R. Holst, *Content Analysis for the Social Sciences and Humanities*. Reading, MA: Perseus Publishing, 1969.

[59] I. Hsieh-Yee, "Research on Web search behavior," *Library & Information Science Research*, vol. 23, pp. 168–185, 2001.

[60] C.-K. Huang, L.-F. Chien, and Y.-J. Oyang, "Relevant term suggestion in interactive web search based on contextual information in query session logs," *Journal of the American Society for Information Science and Technology*, vol. 54, pp. 638–649, 2003.

[61] M.-H. Huang, "Designing Website attributes to induce experiential encounters," *Computers in Human Behavior*, vol. 19, pp. 425–442, 2003.

[62] E. K. R. E. Huizingh, "The antecedents of Web site performance," *European Journal of Marketing*, vol. 36, pp. 1225–1247, 2002.

[63] M. Jackson, "Analytics: Deciphering the Data," 22 January 2007, retrieved 13 May 2009 from http://www.ecommerce-guide.com/resources/article.php/3655251.

[64] B. J. Jansen, "The Methodology of Search Log Analysis," in *Handbook of Research on Web Log Analysis*, B. J. Jansen, A. Spink, and I. Taksa, Eds. Hershey, PA: IGI, 2008, pp. 100–123.

[65] B. J. Jansen, "Paid search," *IEEE Computer*, vol. 39, pp. 88–90, 2006.

[66] B. J. Jansen, "Search log analysis: What is it; what's been done; how to do it," *Library and Information Science Research*, vol. 28, pp. 407–432, 2006.

[67] B. J. Jansen and M. D. McNeese, "Evaluating the effectiveness of and patterns of interactions with automated searching assistance," *Journal of the American Society for Information Science and Technology*, vol. 56, pp. 1480–1503, 2005.

[68]   B. J. Jansen, T. Mullen, A. Spink, and J. Pedersen, "Automated gathering of Web informa-
       tion: An in-depth examination of agents interacting with search engines," *ACM Transac-
       tions on Internet Technology,* vol. 6, pp. 442–464, 2006.

[69]   B. J. Jansen and U. Pooch, "Web user studies: A review and framework for future work,"
       Journal of the American Society of Information Science and Technology, vol. 52, pp. 235–246,
       2001.

[70]   B. J. Jansen and A. Spink, "An Analysis of Web Information Seeking and Use: Documents
       Retrieved Versus Documents Viewed," in *4th International Conference on Internet Comput-
       ing,* Las Vegas, NV, 2003, pp. 65–69.

[71]   B. J. Jansen and A. Spink, "How are we searching the World Wide Web? A compari-
       son of nine search engine transaction logs," *Information Processing & Management,* vol. 42,
       pp. 248–263, 2005.

[72]   B. J. Jansen, A. Spink, C. Blakely, and S. Koshman, "Web searcher interactions with the
       Dogpile.com meta-search engine," *Journal of the American Society for Information Science and
       Technology,* vol. 58, pp. 1875–1887, 2006.

[73]   B. J. Jansen, A. Spink, and T. Saracevic, "Real life, real users, and real needs: A study
       and analysis of user queries on the Web," *Information Processing & Management,* vol. 36,
       pp. 207–227, 2000.

[74]   B. J. Jansen, I. Taksa, and A. Spink, "Research and Methodological Foundations of Transac-
       tion Log Analysis," in *Handbook of Research on Web Log Analysis,* B. J. Jansen, A. Spink, and
       I. Taksa, Eds. Hershey, PA: IGI, 2008, pp. 1–17.

[75]   K. J. Jansen, K. G. Corley , and B. J. Jansen, "E-Survey Methodology: A Review, Issues, and
       Implications," in *Encyclopedia of Electronic Surveys and Measurements (EESM),* J. D. Baker
       and R. Woods, Eds. Hershey, PA: Idea Group Publishing, 2006, pp. 1–8.

[76]   M. Jeong, H. Oh, and M. Gregoire, "Conceptualizing Web site quality and its consequences
       in the lodging industry," *Hospitality Management,* vol. 22, pp. 161–175, 2003.

[77]   T. Joachims, L. Granka, B. Pan, H. Hembrooke, and G. Gay, "Accurately Interpreting Click-
       through Data as Implicit Feedback," in *28th annual international ACM SIGIR conference on
       research and development in information retrieval,* Salvador, Brazil, 2005, pp. 154–161.

[78]   S. Jones, S. Cunningham, and R. McNab, "Usage Analysis of a Digital Library," in *the Third
       ACM Conference on Digital Libraries,* Pittsburgh, PA, 1998, pp. 293–294.

[79]   A. Kaushik (2006, 13 November). Excellent Analytics Tip #8: Measure the Real Conver-
       sion Rate & 'Opportunity Pie,' from http://www.kaushik.net/avinash/2006/11/excellent-
       analytics-tip-8-measure-the-real-conversion-rate-opportunity-pie.html.

[80]   A. Kaushik, *Web Analytics: An Hour a Day.* Indianapolis, IN: Wiley, 2007.

[81] J. Kay and R. C. Thomas, "Studying long-term system use," *Communications of the ACM*, vol. 38, pp. 61–69, 1995.

[82] H.-R. Kea, R. Kwakkelaarb, Y.-M. Taic, and L.-C. Chen, "Exploring behavior of E-journal users in science and technology: Transaction log analysis of Elsevier's ScienceDirect OnSite in Taiwan," *Library & Information Science Research*, vol. 24, pp. 265–291, 2002.

[83] C. M. Kehoe and J. Pitkow, "Surveying the territory: GVU's Five WWW User Surveys," *The World Wide Web Journal*, vol. 1, pp. 77–84, 1996.

[84] M. Kellar, C. Watters, and M. Shepherd, "A field study characterizing Web-based information seeking tasks," *Journal of the American Society for Information Science and Technology*, vol. 58, pp. 999–1018, 2007.

[85] S. Kiesler and L. S. Sproull, "Response effects in the electronic survey," *Public Opinion Quarterly*, vol. 50, pp. 402–413, 1986.

[86] S. Kim and L. Stoel, "Apparel Retailers: Website Quality Dimensions and Satisfaction," *Journal of Retailing and Consumer Services*, vol. 11, pp. 109–117, 2004.

[87] J. Kinsella and P. Bryant, "Online public access catalogue research in the United Kingdom: An overview," *Library Trends*, vol. 35, pp. 619–629, 1987.

[88] R. Korfhage, *Information Storage and Retrieval*. New York: Wiley, 1997.

[89] M. Koufaris, "Applying the technology acceptance model and flow theory to online consumer behavior," *Information Systems Research*, vol. 13, pp. 205–223, 2002.

[90] J. A. Krosnick, "Survey research," *Annual Review of Psychology*, vol. 50, 1999, pp. 537–367.

[91] M. Kurth, "The limits and limitations of transaction log analysis," *Library Hi Tech*, vol. 11, pp. 98–104, 1993.

[92] R. Lempel and S. Moran, "Predictive Caching and Prefetching of Query Results in Search Engines," in *12th international conference on World Wide Web*, Budapest, Hungary, 2003, pp. 19–28.

[93] M. Levene, *An Introduction to Search Engines and Web Navigation*. Essex: Pearson Education, 2005.

[94] S. D. Levitt and J. A. List, "What do laboratory experiments measuring social preferences reveal about the real world?," *Journal of Economic Perspectives*, vol. 21, pp. 153–174, 2007.

[95] S.-J. Lin, "Design Space of Personalized Indexing: Enhancing Successive Web Searching for Transmuting Information Problems," in *Eighth Americas Conference on Information Systems*, Dallas, TX, 2002, pp. 1092–1100.

[96] E. Loken, F. Radlinski, V. H. Crespi, J. Millet, and L. Cushing, "Online study behavior of 100,000 students preparing for the SAT, ACT, AND GRE," *Journal of Educational Computing Research*, vol. 30, pp. 255–262, 2004.

[97] MarketingSherpa, "Security Logo in Email Lifts Average Order Value 28.3%," 20 October 2007, retrieved 13 May 2009 from https://www.marketingsherpa.com/barrier.html?ident=30183.

[98] K. Markey, "Twenty-five years of end-user searching, part 1: Research findings," *Journal of the American Society for Information Science and Technology*, vol. 58, pp. 1071–1081, 2007.

[99] K. Markey, "Twenty-five years of end-user searching, part 2: Future research directions," *Journal of the American Society for Information Science and Technology*, vol. 58, pp. 1123–1130, 2007.

[100] N. Mason, "Customer Loyalty Improves Retention," 6 February 2007, retrieved 30 May 2009 from http://www.clickz.com/showPage.html?page=3624868.

[101] C. McFadden, "Optimizing the Online Business Channel with Web Analytics," 6 July 2005, retrieved 12 May 2009 from http://www.Webanalyticsassociation.org/en/art/?9.

[102] J. E. McGrath, "Methodology Matters: Doing Research in the Behavioral and Social Sciences," in *Readings in Human–Computer Interaction: An Interdisciplinary Approach*, 2nd ed., R. Baecker and W. A. S. Buxton, Eds. San Mateo, CA: Morgan Kaufman Publishers, 1994, pp. 152–169.

[103] V. McKinney, K. Yoon, and F. Zahedi, "The measurement of web-customer satisfaction: An expectation, and disconfirmation approach," *Information Systems Research*, vol. 13, pp. 296–315, 2002.

[104] R. Mehta and E. Sivadas, "Comparing response rates and response content in mail vs. electronic mail surveys," *Journal of the Market Research Society*, vol. 37, pp. 429–439, 1995.

[105] D. Meister and D. Sullivan, "Evaluation of user reactions to a prototype on-line information retrieval system: Report to NASA by the Bunker-Ramo Corporation. Report Number NASA CR-918," Bunker-Ramo Corporation, Oak Brook, IL, 1967.

[106] M. Miwa, "User Situations and Multiple Levels of Users Goals in Information Problem Solving Processes of AskERIC Users," in *2001 Annual Meeting of the American Society for Information Sciences and Technology*, San Francisco, CA, 2001, pp. 355–371.

[107] A. Montgomery and C. Faloutsos, "Identifying Web browsing trends and patterns," *IEEE Computer*, vol. 34, pp. 94–95, July 2001.

[108] D. Nicholas, P. Huntington, N. Lievesley, and R. Withey, "Cracking the code: web log analysis," *Online and CD ROM Review*, vol. 23, pp. 263–269, 1999.

[109] S. Özmutlu and F. Cavdur, "Neural network applications for automatic new topic identification," *Online Information Review*, vol. 29, pp. 34–53, 2005.

[110] S. Özmutlu, H. C. Özmutlu, and A. Spink, "A Study of Multitasking Web Searching," in *the IEEE ITCC'03: International Conference on Information Technology: Coding and Computing*, Las Vegas, NV, 2003, pp. 145–150.

[111] S. Özmutlu, H. C. Özmutlu, and A. Spink, "Topic Analysis and Identification of Queries," in *Handbook of Research on Web Log Analysis*, B. J. Jansen, A. Spink, and I. Taksa, Eds. Hershey, PA: IGI, 2008, pp. 345–358.

[112] S. Page, "Community research: The lost art of unobtrusive methods," *Journal of Applied Social Psychology*, vol. 30, pp. 2126–2136, 2000.

[113] A. Parasuraman, D. Grewal, and R. Krishnan, *Marketing Research*, 2nd ed. Cincinnati, OH: South-Western College Publishing, 1991.

[114] S. Park, H. Bae, and J. Lee, "End user searching: A web log analysis of NAVER, a Korean web search engine," *Library & Information Science Research*, vol. 27, pp. 203–221, 2005.

[115] W. D. Penniman, "Historic Perspective of Log Analysis," in *Handbook of Research on Web Log Analysis*, B. J. Jansen, A. Spink, and I. Taksa, Eds. Hershey, PA: IGI, 2008, pp. 18–38.

[116] W. D. Penniman, "A Stochastic Process Analysis of Online User Behavior," in *Annual Meeting of the American Society for Information Science*, Washington, DC, 1975, pp. 147–148.

[117] T. Peters, "The history and development of transaction log analysis," *Library Hi Tech*, vol. 42, pp. 41–66, 1993.

[118] E. Peterson, *Web Analytics Demystified: A Marketer's Guide to Understanding How Your Web Site Affects Your Business*, New York: Celilo Group Media, 2004.

[119] E. T. Peterson, "Average Order Value," 30 July 2005, retrieved 15 May 2009 from http://blog.webanalyticsdemystified.com/weblog/2005/07/average-order-value.html.

[120] A. Phippen, L. Sheppard, and S. Furnell, "A practical evaluation of Web analytics," *Internet Research: Electronic Networking Applications and Policy*, vol. 14, pp. 284–293, 2004.

[121] J. E. Pitkow, "In Search of Reliable Usage Data on the WWW," in the *Sixth International World Wide Web Conference*, Santa Clara, CA, 1997, pp. 1343–1355.

[122] H. T. Pu, "An exploratory analysis on search terms of network users in Taiwan [in Chinese]," *Central Library Bulletin*, vol. 89, pp. 23–37, 2000.

[123] QuestionPro, "Measuring Customer Loyalty and Customer Satisfaction," n.d., retrieved 15 May 2009 from http://www.questionpro.com/akira/showArticle.do?articleID=customer-loyalty.

[124] L. Rainie and B. J. Jansen, "Surveys as a Complementary Method to Web Log Analysis," in *Handbook of Research on Web Log Analysis*, B. J. Jansen, A. Spink, and I. Taksa, Eds. Hershey, PA: IGI, 2008, pp. 1–17.

[125] L. Rainie and B. J. Jansen, "Surveys as a Complementary Method to Web Log Analysis," in *Handbook of Research on Web Log Analysis*, B. J. Jansen, A. Spink, and I. Taksa, Eds. Hershey, PA: IGI, 2008, pp. 39–64.

[126] R. E. Rice and C. L. Borgman, "The use of computer-monitored data in information science," *Journal of the American Society for Information Science*, vol. 44, pp. 247–256, 1983.

[127] S. Y. Rieh and H. Xu, "Patterns and Sequences of Multiple Query Reformulation in Web Searching: A Preliminary Study," in *64th Annual Meeting of the American Society for Information Science and Technology*, 2001, pp. 246–255.

[128] N. C. Romano, C. Donovan, H. Chen, and J. F. Nunamaker, "A methodology for analyzing Web-based qualitative data," *Journal of Management Information Systems*, vol. 19, pp. 213–246, 2003.

[129] D. E. Rose and D. Levinson, "Understanding User Goals in Web Search," in *World Wide Web Conference (WWW 2004)*, New York, 2004, pp. 13–19.

[130] G. M. Rose, M. L. Meuter, and J. L. Curran, "On-line waiting: The role of download time and other important predictors on attitude toward e-retailers," *Psychology & Marketing Research*, vol. 22, pp. 127–151, 2005.

[131] N. Ross and D. Wolfram, "End user searching on the Internet: an analysis of term pair topics submitted to the excite search engine," *Journal of the American Society for Information Science,* vol. 51, pp. 949–958, 2000.

[132] D. Sapir, "Online Analytics and Business Performance Management," in *BI Report*, 2004.

[133] SearchCRM, "Measuring Customer Loyalty," 9 May 2007, retrieved 15 May 2009 from http://searchcrm.techtarget.com/general/0,295582,sid11_gci1253794,00.html.

[134] W. Sellars, "Philosophy and the Scientific Image of Man," in *Science, Perception, and Reality.* New York: Ridgeview Publishing, 1963, pp. 1–40.

[135] C. Silverstein, M. Henzinger, H. Marais, and M. Moricz, "Analysis of a Very Large Web Search Engine Query Log," *SIGIR Forum*, vol. 33, pp. 6–12, 1999.

[136] F. Silvestri, "Mining query logs: Turning search usage data into knowledge," *Foundations and Trends in Information Retrieval*, in press.

[137] B. F. Skinner, *Science and Human Behavior.* New York: Free Press, 1953.

[138] A. Spink, "Multitasking information behavior and information task switching: An exploratory study," *Journal of Documentation*, vol. 60, pp. 336–345, 2004.

[139] A. Spink, J. Bateman, and B. J. Jansen, "Searching the Web: A survey of Excite users," *Journal of Internet Research: Electronic Networking Applications and Policy*, vol. 9, pp. 117–128, 1999.

[140] A. Spink and B. J. Jansen, *Web Search: Public Searching of the Web*. New York: Kluwer, 2004.

[141] A. Spink, T. Wilson, D. Ellis, and F. Ford. (1998, April 1998). Modeling Users' Successive Searches in Digital Environments. D-Lib Magazine.

[142] L. S. Sproull, "Using electronic mail for data collection in organizational research," *Academy of Management Journal*, vol. 29, pp. 159–169, 1986.

[143] J. Sterne, "10 Steps to Measuring Website Success," n.d., retrieved 15 May 2009 from http://www.marketingprofs.com/login/join.asp?adref=rdblk&source=/4/sterne13.asp.

[144] J. Sterne, "Web Channel Performance Management: Aligning Web Site Vision and Strategy with Goals and Tactics," Pilot Software Inc., Mountain View, CA, 2004.

[145] S. Sudman, N. M. Bradburn, and N. Schwarz, *Thinking about answers: The application of cognitive processes to survey methodology.* San Francisco: Jossey-Bass Publishers, 1996.

[146] D. Sullivan, "SpiderSpotting: When A Search Engine, Robot Or Crawler Visits," 6 November 2001, retrieved 5 August 2003 from http://www.searchenginewatch.com/web masters/article.php/2168001.

[147] M. Thelwall, *Introduction to Webometrics: Quantitative Web Research for the Social Sciences.* San Rafael, CA: Morgan and Claypool, 2009.

[148] H. Treiblmaier, "Web site analysis: A review and assessment of previous research," *Communications of the Association for Information Systems,* vol. 19, pp. 806–843, 2007.

[149] A. C. B. Tse, K. C. Tse, C. H. Yin, C. B. Ting, K. W. Yi, K. P. Yee, and W. C. Hong, "Comparing two methods of sending out questionnaires: E-mail vs. mail," *Journal of the Market Research Society,* vol. 37, pp. 441–446, 1995.

[150] K. Waite and T. Harrison, "Consumer expectations of online information provided by bank websites," *Journal of Financial Services Marketing* vol. 6, pp. 309–322, 2002.

[151] P. Wang, M. Berry, and Y. Yang, "Mining longitudinal web queries: Trends and patterns," *Journal of the American Society for Information Science and Technology,* vol. 54, pp. 743–758, 2003.

[152] J. B. Watson, "Psychology as the behaviorist views it," *Psychological Review,* vol. 20, pp. 158–177, 1913.

[153] E. J. Webb, D. T. Campbell, R. D. D. Schwartz, L. Sechrest, and J. B. Grove, *Nonreactive measures in the social sciences,* 2nd ed. ed. Boston, MA: Houghton Mifflin, 1981.

[154] E. J. Webb, D. T. Campbell, R. D. Schwarz, and L. Sechrest, *Unobtrusive Measures (Revised Edition).* Thousand Oaks, CA: Sage, 2000.

[155] WebSideStory, "Use of Key Performance Indicators in Web Analytics," 20 May 2004, retrieved 10 May 2009 from www.4everywhere.com/documents/KPI.pdf.

[156] J.-R. Wen, J.-Y. Nie, and H.-J. Zhang, "Clustering User Queries of a Search Engine," in *10th International Conference on World Wide Web,* Hong Kong, 2001, pp. 162–168.

[157] K. White, "Unique vs. Returning Visitors Analyzed," 10 May 2006, retrieved 31 May 2009 from http://newsletter.blizzardinternet.com/unique-vs-returning-visitors-analyzed/2006/05/10/#more-532.

[158] D. Wolfram, "Term co-occurrence in internet search engine queries: An analysis of the Excite data set," *Canadian Journal of Information and Library Science,* vol. 24, pp. 12–33, 1999.

[159]   D. Wolfram, P. Wang, and J. Zhang, "Identifying Web search session patterns using cluster analysis: A comparison of three search environments," *Journal of the American Society for Information Science and Technology*, vol. 60, pp. 896–910, 2009.

[160]   Y. Xie and D. O'Hallaron, "Locality in Search Engine Queries and Its Implications for Caching," in *the Twenty-First Annual Joint Conference of the IEEE Computer and Communications Societies*, New York, 2002, pp. 307–317.

[161]   D. Young, "Site search: increases conversion rates, average order value and loyalty," *Practical Ecommerce*, vol. 2009, 2007.

[162]   L. Yu and A. Apps, "Studying e-journal user behavior using log files: The experience of SuperJournal," *Library & Information Science Research*, vol. 22, pp. 311–338, 2000.

[163]   W. Yuan and C. T. Meadow, "A study of the use of variables in information retrieval user studies," *Journal of the American Society for Information Science*, vol. 50, pp. 140–150, 1999.

[164]   G. K. Zipf, *Human Behavior and the Principle of Least Effort*. Cambridge, MA: Addison-Wesley Press, 1949.

Author Biography

Bernard J. Jansen is an associate professor in the College of Information Sciences and Technology at the Pennsylvania State University. Jim has more than 150 publications in the area of information technology and systems, with articles appearing in a multidisciplinary range of journals and conferences. His specific areas of expertise are Web searching, sponsored search, and personalization for information searching. He is coauthor of the book, *Web Search: Public Searching of the Web* and coeditor of the book *Handbook of Weblog Analysis*. Jim is a member of the editorial boards of eight international journals. He has received several awards and honors, including an ACM Research Award and six application development awards, along with other writing, publishing, research, and leadership honors. Several agencies and corporations have supported his research. He is actively involved in teaching both undergraduate and graduate level courses, as well as mentoring students in a variety of research and educational efforts. He also has successfully conducted numerous consulting projects. Jim resides with his family in Charlottesville, VA.